MCQs in Biochemistry

(For All India PG Entrance Examination)

Harbans Lal PhD, FIAO, FACBI
Senior Professor and Head, Department of Biochemistry
Maharaja Agrasen Medical College, Agroha (Hisar)
Former Senior Professor, Department of Biochemistry, PGIMS
Pt BD Sharma University of Health Sciences, Rohtak
Former Visiting Professor, Department of Biosciences
MD University, Rohtak
Former WHO Fellow and Visiting Assistant Professor
Louisiana State University Medical Centre, New Orleans (USA)

Rajesh Pandey MD
Associate Professor, Department of Biochemistry
Maharishi Markandeshwar Institute of Medical Sciences and Research, Mullana, Ambala (Haryana)

CBS Publishers & Distributors Pvt Ltd

New Delhi • Bengaluru • Chennai • Kochi • Kolkata • Mumbai
Hyderabad • Jharkhand • Nagpur • Patna • Pune • Uttarakhand

MCQs in Biochemistry

(For All India PG Entrance Examination)

ISBN: 978-81-239-2305-5

First Edition: 2013

Reprint: 2014, 2016, 2022

Published by **Satish Kumar Jain** and produced by **Varun Jain** for

CBS Publishers & Distributors Pvt Ltd

4819/XI Prahlad Street, 24 Ansari Road, Daryaganj, New Delhi 110 002, India.

Ph: 011-23289259, 23266861, 23266867 Website: www.cbspd.com

Fax: 011-23243014 e-mail: delhi@cbspd.com; cbspubs@airtelmail.in.

Corporate Office: 204 FIE, Industrial Area, Patparganj, Delhi 110 092

Ph: 011-4934 4934 Fax: 011-4934 4935

e-mail: publishing@cbspd.com; publicity@cbspd.com

Branches

- **Bengaluru:** Seema House 2975, 17th Cross, K.R. Road, Banasankari 2nd Stage, Bengaluru 560 070, Karnataka, India
  Ph: +91-80-26771678/79 Fax: +91-80-26771680 e-mail: bangalore@cbspd.com
- **Chennai:** 7, Subbaraya Street, Shenoy Nagar, Chennai 600 030, Tamil Nadu, India
  Ph: +91-44-26680620, 26681266 Fax: +91-44-42032115 e-mail: chennai@cbspd.com
- **Kochi:** 42/1325, 1326, Power House Road, Opp KSEB, Ernakulum, Kochi 682 018, Kerala, India
  Ph: +91-484-4059061-65,67 Fax: +91-484-4059065 e-mail: kochi@cbspd.com
- **Kolkata:** 147, Hind Ceramics Compound, 1st Floor, Nilgunj Road, Belghoria, Kolkata-700056, India
  Ph: +91-9096713055/7798394118, 9836841399 e-mail: kolkata@cbspd.com
- **Mumbai:** PWD Shed, Gala no 25/26, Ramchandra Bhatt Marg, Next to JJ Hospital Gate no. 2, Opp. Union Bank of India, Noorbaug Mumbai-400009, Maharashtra, India
  Ph: +91-22-66661880/89 e-mail: mumbai@cbspd.com

Representatives

- Hyderabad 0-9885175004
- Jharkhand 0-9811541605
- Nagpur 0-9421945513
- Patna 0-9334159340
- Pune 0-9623451994
- Uttarakhand 0-9716462459

Printed at India Binding House, Noida, UP, India

Preface

The present compilation, in the subject of Biochemistry, endeavors us to help students preparing for the All India PG Medical Entrance Examination. It includes questions from all categories, i.e. from the simplest ones to real brain teasers, in order to cater the needs of the unprepared students who start studying at the eleventh hour, besides, of course, the committed ones who are well conversant with the subject. As teachers, we have experienced that because Biochemistry is taught in the first year of the MBBS curriculum and that by the time a student starts preparing for the postgraduate entrance examination, he/she has forgotten most of its contents. Moreover, significant advances made in the field of molecular biology including molecular aspects of disease processes, immunology, cancer, etc., are not suitably covered during the MBBS curriculum, may be due to time strains. Many MCQs are based on such latest advances. We have also incorporated numerous MCQs that are no doubt linked to Biochemistry but nevertheless in an interdisciplinary sense. We feel that, if read along with any standard text book of Biochemistry, the MCQs should strengthen the overall preparation process. We hope that the book will be welcome by the students and the teachers alike. Constructive criticisms will help us to improve the future edition of the book.

Harbans Lal
(hl.biopgimsr@gmail.com)
Rajesh Pandey
(calldrrajesh@yahoo.co.in)

Contents

SECTION-1

CELL, BIOMEMBRANES, EXTRACELLULAR MATRIX

1. Nucleoid is found in:

(a) Animal cell (b) Plant cell
(c) Bacterium (d) Virus

2. The following is not a true organelle:

(a) Ribosome (b) Mitochondrion
(c) Lysosome (d) Nucleus

3. Isolation of the following subcellular entity requires ultracentrifugation:

(a) Endoplasmic reticulum (b) Lysosome
(c) Nucleus (d) Peroxisome

4. Galactosyl transferase is a biomarker for the following subcellular entity:

(a) Cell membrane (b) Cytoplasm
(c) Golgi apparatus (d) Endoplasmic reticulum

5. Glucose-6-phosphatase is a biomarker for the following subcellular entity:

(a) Cell membrane (b) Cytoplasm
(c) Golgi apparatus (d) Endoplasmic reticulum

6. The following structure disappears during cell division:

(a) Nuclear membrane (b) Chromosomes
(c) Nucleoplasm (d) Nucleolus

7. Protein synthesis occurs in:

(a) Ribosomes only
(b) Mitochondria only
(c) Ribosomes and mitochondria
(d) Smooth endoplasmic reticulum and mitochondria

8. Protein sorting occurs in:

(a) Golgi apparatus
(b) Endoplasmic reticulum
(c) Cell membrane
(d) Mitochondria

9. Power house of a cell is:

(a) Mitochondria
(b) Lysosomes
(c) Nucleus
(d) Ribosomes

10. The mitochondrial DNA encodes:

(a) 13 polypeptides
(b) 13 polypeptides and 2rRNAs
(c) 13 polypeptides, 2rRNAs and 22tRNAs
(d) 2rRNAs and 22tRNAs

11. In contrast to the outer mitochondrial membrane, the inner mitochondrial membrane has:

(a) More lipids than proteins
(b) More proteins than lipids
(c) Only proteins
(d) Only lipids

12. Mitochondrial DNA is:

(a) Paternally inherited and more prone to mutations than nuclear DNA
(b) Paternally inherited and less prone to mutations than nuclear DNA
(c) Maternally inherited and more prone to mutations than nuclear DNA
(d) Maternally inherited and less prone to mutations than nuclear DNA

13. Membrane folds, called cristae, are associated with:

(a) Outer nuclear membrane
(b) Inner nuclear membrane

(c) Outer mitochondrial membrane
(d) Inner mitochondrial membrane

14. 'Suicide bags' refer to:

(a) Lysosomes (b) Peroxisomes
(c) Ribosomes (d) Nucleus

15. The following enzyme is missing in cholesterol ester storage disease:

(a) Lysosomal acid lipase (b) LCAT
(c) ACAT (d) HMG CoA lyase

16. Zellweger syndrome is due to the absence or loss of functions of:

(a) Lysosomes (b) Peroxisomes
(c) Ribosomes (d) Endoplasmic reticulum

17. The ideal source for obtaining stem cells is:

(a) Adult peripheral blood
(b) Adult bone marrow
(c) Neonatal peripheral blood
(d) Umbilical cord blood

18. Till date, the major success of stem cell therapy is limited to:

(a) Leukemia (b) AIDS
(c) Alzheimer's disease (d) Muscular dystrophy

19. When the cross-section of the hydrophilic head group of amphipathic lipids is more than that of the hydrocarbon tails, the following are formed:

(a) Liposomes (b) Micelles
(c) Lipid bilayers (d) Lipid monolayers

20. Liposomes can be used to deliver:

(a) DNA (b) Enzymes
(c) Amphotericin B (d) All the above

21. To avoid non-specific uptake of liposomes by the reticulo-endothelial system, they are coated with:

(a) Polyethylene glycol (b) Cholesterol
(c) Sphingolipids (d) Cations

22. Liposomes can be used to deliver:

(a) Hydrophilic compounds (b) Lipophilic compounds
(c) Any of the above (d) None of the above

23. In the myelin sheath, lipid:protein ratio is nearly:

(a) 4:1 (b) 1:4
(c) 1:1 (d) 3:2

24. In the inner mitochondrial membrane, lipid:protein ratio is nearly:

(a) 4:1 (b) 1:4
(c) 1:1 (d) 3:2

25. The following is not true for Glycophorin A:

(a) Present in RBC membrane
(b) Peripheral membrane protein
(c) Tightly bound by hydrophobic interactions
(d) Can be isolated by disrupting the membrane with detergents

26. The following is not true for cytochrome c:

(a) Integral membrane protein
(b) Present in inner mitochondrial membrane
(c) Can be isolated by gentle procedures
(d) Soluble component of electron transport chain

27. In biomembranes, oligosaccharides may be attached to:

(a) Protein (b) Lipid
(c) Either of the above (d) None of the above

28. Peripheral membrane proteins may be attached to:

(a) Integral membrane proteins
(b) Phospholipid head groups
(c) Glycosylphosphatidyl inositol
(d) Any of the above

29. Membrane fluidity depends upon:

(a) Its cholesterol content
(b) Its unsaturated fatty acid content

(c) Degree of unsaturation of fatty acids
(d) All the above

30. The following is not true for cholesterol that is present in biomembranes:

(a) Always esterified
(b) Steroid nucleus intercalates between phospholipid hydrocarbon tails
(c) Hydrocarbon chain embedded in non-polar membrane core
(d) OH-group interacts with the polar head group of the phospholipid

31. Sodium-glucose transporter-1 (SGLT-1) is present in:

(a) Apical membrane of proximal convoluted tubules
(b) Intestinal brush border
(c) Both
(d) None of the above

32. In myasthenia gravis, autoantibodies are directed against the following membrane receptor:

(a) Dopamine (b) Epinephrine
(c) Acetyl choline (d) Serotonin

33. The following treatment modality is not employed for myasthenia gravis:

(a) Plasmapheresis
(b) Acetyl cholinesterase
(c) Acetyl cholinesterase inhibitors
(d) None of the above

34. Cystic fibrosis transmembrane conductance regulator (CFTR) is a:

(a) cGMP-dependent Ca^{2+}-channel
(b) cGMP-dependent Cl^--channel
(c) cAMP-dependent Ca^{2+}-channel
(d) cAMP-dependent Cl^--channel

35. The following is true for aquaporins:

(a) Have low molecular weight
(b) Integral membrane proteins

(c) Responsible for rapid transmembrane water movement
(d) All the above

36. In the renal collecting tubules, expression of aquaporin-2 (AQP2) is increased by:
(a) ADH (b) ANP
(c) Aldosterone (d) Angiotensin II

37. In renal collecting tubules, following drug may decrease the expression of aquaporins:
(a) Aceclofenac (b) Aminoglycosides
(c) Cisplatin (d) Lithium

38. The following is not true for diabetes insipidus:
(a) Increased urine specific gravity
(b) Polyuria
(c) Decreased expression of aquaporins in collecting tubules
(d) Treated with arginine-vasopressin

39. The following is an ionophore:
(a) Valinomycin (b) Nigericin
(c) Both (d) None of the above

40. Valinomycin and nigericin translocate:
(a) K^+ (b) Na^+
(c) Ca^{2+} (d) Cl^-

41. ATP-ADP translocase is present in:
(a) Cell membrane
(b) Nuclear membrane
(c) Inner mitochondrial membrane
(d) Lysosomal membrane

42. F-type transporters are present in:
(a) Lysosomal membrane (b) Mitochondrial membrane
(c) Nuclear membrane (d) Plasma membrane

43. V-type transporters are not present in:
(a) Lysosomal membrane (b) Mitochondrial membrane
(c) Golgi membrane (d) Secretory vesicles

44. The function of lysosomal V-type transporters is:

(a) Acidification of the interior of lysosomes
(b) Alkalinization of the interior of lysosomes
(c) ATP synthesis
(d) Promote membrane cleavage

45. Na^+-K^+-ATPase is inhibited by:

(a) Digoxin
(b) Ouabain
(c) Both
(d) None of the above

46. Cardiac glycosides act by:

(a) Increasing intracellular Ca^{2+}
(b) Decreasing intracellular Ca^{2+}
(c) Decreasing intracellular Na^+
(d) Increasing intracellular K^+

47. Gastric proton pump is:

(a) A uniport
(b) A symport
(c) An electrogenic antiport
(d) A neutral antiport

48. A defect in membrane serotonin transporter is implicated in:

(a) Autism
(b) Gastric motility disorders
(c) Obsessive compulsive disorders
(d) All the above

49. A defect in membrane glutamate transporter is implicated in:

(a) Amyotrophic lateral sclerosis
(b) Epilepsy
(c) Schizophrenia
(d) All the above

50. Oral rehydration solution (ORS) therapy exploits the following transport system in the small intestine:

(a) GLUT
(b) SGLT-1

(c) AQP
(d) Neutral amino acid transporter

51. In group translocation, for each amino acid transported, the number of ATP required is:

(a) 0 (b) 1
(c) 2 (d) 3

52. The following is not true for adsorptive pinocytosis:

(a) Receptor-mediated
(b) Clathrin-dependent
(c) No secondary lysosomes are formed
(d) Explains the mechanism of action of some hormones

53. The following leave a cell by exocytosis except:

(a) miRNA (b) Hormones
(c) Blood group antigens (d) Collagens

54. Exosomes contain:

(a) RNase (b) mRNA
(c) miRNA (d) All the above

55. The most abundant protein in human body is:

(a) Elastin (b) Fibrillin
(c) Collagen (d) Fibronectin

56. Collagens assume a gel-like consistency in:

(a) Vitreous humor (b) Tendons
(c) Bones (d) Cornea

57. The following is not true for collagens:

(a) Fibrous proteins (b) Extracellular
(c) Glycoproteins (d) Lack quaternary structure

58. Basement membranes mainly contain collagen type:

(a) III (b) IV
(c) VII (d) XII

59. Tendons and ligaments mainly contain collagen type:

(a) III (b) IV
(c) VII (d) XII

60. Fetal skin mainly contains collagen type:

(a) III (b) IV
(c) VII (d) XII

61. The commonest amino acid in collagens is:

(a) Glycine (b) Proline
(c) Hydroxyproline (d) Lysine

62. Collagens are synthesized by:

(a) Osteoblasts (b) Chondroblasts
(c) Fibroblasts (d) All the above

63. The formation of interchain H-bonds in collagens depends on the presence of:

(a) Glycine (b) Proline
(c) Hydroxyproline (d) Lysine

64. The formation of interchain covalent bonds in collagens depends on the presence of:

(a) Glycine (b) Proline
(c) Hydroxyproline (d) Hydroxylysine

65. Lysyl oxidase requires:

(a) Oxygen (b) Cupric ions
(c) Ascorbic acid (d) All the above

66. Lysyl hydroxylase requires:

(a) Oxygen (b) Ferrous ions
(c) Ascorbic acid (d) All the above

67. The following is true for extension peptides in procollagen:

(a) Completely retained in mature collagens
(b) Removed intracellularly
(c) Removed extracellularly
(d) Partly retained in mature collagens

68. Osteogenesis imperfecta involves collagen type:

(a) I (b) II
(c) III (d) IV

69. **A common cause of death in type II osteogenesis imperfecta is:**
(a) Fractures (b) Epistaxis
(c) Hematuria (d) Pulmonary hypoplasia

70. **Desmosine cross-links are found in:**
(a) Collagens (b) Elastin
(c) Fibrillin (d) Fibronectin

71. **In elastin, every one in seven amino acids is:**
(a) Glycine (b) Lysine
(c) Proline (d) Valine

72. **In collagens, every one in three amino acids is:**
(a) Glycine (b) Lysine
(c) Proline (d) Valine

73. **The elasticity of elastin is due to:**
(a) Desmosine cross-links (b) Valine-rich domains
(c) Both (d) None of the above

74. **The following is true for Marfan's syndrome:**
(a) Floppy mitral valves (b) Arachnodactyly
(c) Fibrillin gene mutation (d) All the above

75. **The following is not true for keratan sulphate:**
(a) Most homogeneous group of glycosaminoglycans
(b) Keratan sulphate-I is found in cornea
(c) May be N-linked or O-linked with proteins
(d) Sulphate content is variable

76. **Epimerization at C5 of glucuronate forms:**
(a) Iduronate (b) Gulonate
(c) Guluronate (d) Mannuronate

77. **The enzyme deficient in Hunter syndrome is:**
(a) Iduronate sulphatase (b) α-L-Iduronidase
(c) Galactose-6-sulphatase (d) β-Galactosidase

78. The enzyme deficient in Hurler syndrome is:

(a) Iduronate sulphatase (b) α-L-Iduronidase
(c) Galactose-6-sulphatase (d) β-Galactosidase

79. The enzyme deficient in Morquio syndrome 'A' is:

(a) Iduronate sulphatase (b) α-L-Iduronidase
(c) Galactose-6-sulphatase (d) β-Galactosidase

80. The enzyme deficient in Morquio syndrome 'B' is:

(a) Iduronate sulphatase (b) α-L-Iduronidase
(c) Galactose-6-sulphatase (d) β-Galactosidase

81. The enzyme deficient in Sanfilippo syndrome 'A' is:

(a) Heparan sulphamidase
(b) N-Acetyl glycosaminidase
(c) Acetyl CoA: glucosaminide N-acetyl transferase
(d) N-Acetyl glucosamine-6-sulphate sulphatase

82. The enzyme deficient in Sanfilippo syndrome 'B' is:

(a) Heparan sulphamidase
(b) N-Acetyl glycosaminidase
(c) Acetyl CoA: glucosaminide N-acetyl transferase
(d) N-Acetyl glucosamine-6-sulphate sulphatase

83. The enzyme deficient in Sanfilippo syndrome 'C' is:

(a) Heparan sulphamidase
(b) N-Acetyl glycosaminidase
(c) Acetyl CoA: glucosaminide N-acetyl transferase
(d) N-Acetyl glucosamine-6-sulphate sulphatase

84. The enzyme deficient in Sanfilippo syndrome 'D' is:

(a) Heparan sulphamidase
(b) N-Acetyl glycosaminidase
(c) Acetyl CoA: glucosaminide N-acetyl transferase
(d) N-Acetyl glucosamine-6-sulphate sulphatase

85. The enzyme deficient in Sly syndrome is:

(a) β-Glucuronidase (b) β-Glucosidase
(c) β-Galactosidase (d) β-Galacturonidase

86. The following is/are podocyte cell surface protein(s)

(a) Nephrin
(b) Podocin
(c) Both
(d) None of the above

87. The following is not a feature of Liddle's syndrome:

(a) Abnormal apical membrane Na^+-channel in renal cortical collecting ducts
(b) Hypotension
(c) Hypokalemia
(d) Metabolic alkalosis

88. Skeletal muscle membrane Na^+-K^+-ATPase is stimulated by:

(a) Insulin
(b) Epinephrine
(c) Both
(d) None of the above

89. Digitalis overdose causes hyperkalemia by:

(a) Inhibiting muscle Na^+-K^+-ATPase
(b) Stimulating muscle Na^+-K^+-ATPase
(c) Inhibiting liver Na^+-K^+-ATPase
(d) Stimulating liver Na^+-K^+-ATPase

90. The following is an 'adaptor protein':

(a) Spectrin
(b) Ankyrin
(c) Actin
(d) Calmodulin

91. Hereditary spherocytosis is due to defective:

(a) Ankyrin R
(b) Ankyrin B
(c) Ankyrin G
(d) None of the above

92. Spectrin and ankyrin function together to:

(a) Maintain RBC shape
(b) Expression of antigens on T lymphocytes
(c) Assembly of axonal membrane ion transporters
(d) All the above

93. Defect in RBC membrane glycosylphosphatidyl inositol results in:

(a) Paroxysmal nocturnal hemoglobinuria
(b) Sickle cell anemia

(c) α-Thalassemia
(d) β-Thalassemia

94. Extracellular fibrillar amyloid deposits in Alzheimer's disease are rich in:

(a) Amylin
(b) Amyloid β-peptide (Aβ42)
(c) Amyloid precursor protein
(d) α-Synuclein

95. The following is true for flip-flop movement of membrane phospholipids:

(a) Catalyzed by flippases
(b) Move phospholipids from one leaflet to the other
(c) Establish lipid asymmetry
(d) All the above

96. The following types of movements of phospholipids occur in biomembranes:

(a) Flexion
(b) Lateral shift
(c) Transverse diffusion (flip-flop)
(d) All the above

97. Compared to the surrounding plasma membrane, lipid rafts are rich in:

(a) Sphingolipids
(b) Cholesterol
(c) Both
(d) None of the above

98. The following is true for lipid rafts:

(a) Planar rafts (non-caveolar) contain the protein flotillin
(b) Caveolar rafts contain the protein caveolin
(c) Employed by influenza viruses and HIV-1 for infecting host cells
(d) All the above

99. Caveolins are:

(a) Integral membrane proteins
(b) Peripheral membrane proteins
(c) Cytosolic proteins
(d) Extracellular matrix (ECM) proteins

100. Freeze-fracture replication is used to study:

(a) DNA replication under cold conditions
(b) Integral membrane proteins
(c) Reverse transcription
(d) Transmembrane potential difference

SECTION-2

CHEMISTRY AND METABOLISM OF CARBOHYDRATES, LIPIDS, AMINO ACIDS, NUCLEOTIDES

1. A chiral center refers to:

(a) A symmetric carbon atom with 4 identical atoms or groups attached to it

(b) A symmetric carbon atom with 4 nonidentical atoms or groups attached to it

(c) An asymmetric carbon atom with 4 nonidentical atoms or groups attached to it

(d) An asymmetric carbon atom with 4 identical atoms or groups attached to it

2. Two enantiomers rotate the plane of polarized light:

(a) To the same degree in the same direction

(b) To the same degree but in opposite directions

(c) To different degrees in the same direction

(d) To different degrees in opposite directions

3. If the number of chiral centres in a molecule is 5, the number of possible stereoisomers is:

(a) $5 \times 2 = 10$ (b) $5^2 = 25$

(c) $2^5 = 32$ (d) 5

4. A chiral molecule having D-configuration can be:

(a) Only dextrorotatory [D(+)]

(b) Only levorotatory [D(–)]

(c) Dextrorotatory [D(+)] or levorotatory [D(–)]
(d) Dextrorotatory [D(+)] or levorotatory [L(–)]

5. Parent compound of the carbohydrate family is:

(a) Glucose
(b) Glycogen
(c) Galactose
(d) Glyceraldehyde

6. Haworth structures refer to:

(a) Pyran and furan forms of sugars
(b) 'D' and 'L' forms of sugars
(c) 'd' and 'l' forms of sugars
(d) Aldose and ketose forms of sugars

7. The smallest furan monosaccharide is:

(a) Glucose
(b) Mannose
(c) Ribose
(d) Arabinose

8. In a cyclic monosaccharide, the anomeric carbon refers to:

(a) C1 in an aldose and C2 in a ketose
(b) C2 in an aldose and C1 in a ketose
(c) C1 in an aldose or a ketose
(d) C2 in an aldose or a ketose

9. In a glucose solution at equilibrium, mutarotation results in a fixed optical rotation of:

(a) 55.2°
(b) 5.52°
(c) 5.25°
(d) 52.5°

10. With respect to D-Glucose, D-Mannose and D-Galactose are the epimers at:

(a) C2 and C4, respectively
(b) C4 and C2, respectively
(c) C2 and C1, respectively
(d) C1 and C2, respectively

11. Oxidation of C1 aldehyde group of glucose to an acid forms:

(a) Glucuronic acid
(b) Gluconic acid
(c) Guluronic acid
(d) Glucuronide

12. Oxidation of C6 alcohol group of glucose to an acid forms:

(a) Glucuronic acid (b) Gluconic acid
(c) Guluronic acid (d) Glucuronide

13. Sialic acids (N-acetylneuraminic acid and its derivatives) are important constituents of:

(a) Glycoproteins (b) Nucleoproteins
(c) Lipoproteins (d) Phosphoproteins

14. Drugs used against influenza virus (e.g. swine flu virus) generally target the following viral enzyme:

(a) Endonuclease (b) Integrase
(c) Neuraminidase (d) Helicase

15. An example of a nonreducing disaccharide is:

(a) Maltose (b) Isomaltose
(c) Lactose (d) Sucrose

16. Invert sugar, cane sugar and table sugar are the other names of:

(a) Sucrose (b) Glucose
(c) Fructose (d) Lactose

17. The most notorious cariogenic sugar is:

(a) Sucrose (b) Glucose
(c) Fructose (d) Lactose

18. In the 3rd trimester of pregnancy, urine may normally contain:

(a) Sucrose (b) Glucose
(c) Fructose (d) Lactose

19. Lactulose is a clinically useful semisynthetic disaccharide and is composed of:

(a) Fructose and galactose (b) Fructose and glucose
(c) Glucose and galactose (d) Glucose and glucose

20. Inulin is a:

(a) Glucosan (b) Mannosan
(c) Fructosan (d) Galactosan

21. In humans, the enzymes capable of hydrolyzing β-(1→4) linkages of dietary polysaccharides are found in:

(a) Saliva (b) Pancreatic juice
(c) Intestinal juice (d) Nowhere

22. In humans, the enzymes capable of hydrolyzing α-(1→4) linkages of dietary polysaccharides are found in:

(a) Saliva only
(b) Pancreatic juice only
(c) Saliva and pancreatic juice
(d) Nowhere

23. Amylose and amylopectin are characteristically found in:

(a) Starch (b) Glycogen
(c) Cellulose (d) Inulin

24. In the treatment of hypovolemic shock, the following is used as a plasma volume expander by intravenous infusion:

(a) Limit dextrin (b) Dextrin
(c) Dextrose (d) Dextran

25. In the treatment of hypoglycemic shock, the following is used as an instant energy source by intravenous infusion:

(a) Limit dextrin (b) Dextrin
(c) Dextrose (d) Dextran

26. The following homopolysaccharide is found in the exoskeleton of insects:

(a) Chitin (b) Cellulose
(c) Glycogen (d) Inulin

27. Heteropolysaccharides are abundant in:

(a) Extracellular matrix (b) Blood circulation
(c) Cytosol (d) Intestinal secretions

28. In the prophylaxis and treatment of deep vein thrombosis, low molecular weight derivatives of the following heteropolysaccharide are used:

(a) Hyaluronic acid (b) Chondroitin sulphate
(c) Heparan (d) Heparin

29. The following heteropolysaccharide is commercially available for injecting into synovial joints in patients suffering from advanced osteoarthritis:

(a) Hyaluronic acid
(b) Chondroitin sulphate
(c) Heparan
(d) Heparin

30. Iduronate is present in:

(a) Hyaluronic acid
(b) Chondroitin-4-sulphate
(c) Dermatan sulphate
(d) Chondroitin-6-sulphate

31. Mast cell granules are rich in the following heteropolysaccharide:

(a) Keratan sulphate
(b) Chondroitin-4-sulphate
(c) Dermatan sulphate
(d) Heparin

32. Cornea and heart valves are rich in the following heteropolysaccharide(s):

(a) Keratan sulphate
(b) Chondroitin-4-sulphate and chondroitin-6-sulphate
(c) Dermatan sulphate
(d) Keratan sulphate and dermatan sulphate

33. Vitreous humor contains a heteropolysaccharide having the following repeating disaccharide units:

(a) D-glucuronic acid and N-acetyl galactosamine-4-sulphate
(b) D-glucuronic acid and N-acetyl galactosamine-6-sulphate
(c) D-glucuronic acid and N-acetyl glucosamine
(d) Iduronate and N-acetyl galactosamine

34. Salivary amylase acts on:

(a) 1,4α-glycosidic linkage, randomly
(b) 1,4β-glycosidic linkage, randomly
(c) 1,4α-glycosidic linkage, from terminal end only
(d) 1,4β-glycosidic linkage, from terminal end only

35. The following is not a product of digestion of carbohydrates by pancreatic amylase:

(a) Limit dextrin
(b) D-Glucose

(c) Dextran
(d) None of the above

36. The following is a product of digestion of polysaccharides by pancreatic amylase:
(a) Limit dextrin
(b) Isomaltose
(c) Maltotriose
(d) All the above

37. Isomaltase is present in:
(a) Saliva
(b) Gastric juice
(c) Pancreatic juice
(d) Small intestinal brush border

38. The following is true for sucrase deficiency:
(a) May be congenital
(b) Failure to hydrolyze sucrose in the small intestine
(c) Symptoms similar to lactase deficiency
(d) All the above

39. The following is not true for milk allergy:
(a) Also called lactose intolerance
(b) Gastrointestinal manifestations following milk intake
(c) Immune reaction to milk protein
(d) Seen in newborns

40. Generally, lactase activity is highest:
(a) Immediately after birth
(b) In children
(c) In young adults
(d) In elderly

41. The following is not true for lactose intolerance:
(a) Intolerance to cheese
(b) Symptoms partly due to presence of osmotically active lactose in the colon
(c) Prolonged diarrhea itself may cause secondary lactase deficiency
(d) Symptoms partly due to colonic fermentation of lactose to carbon dioxide and organic acids

42. The following is not true for acidophilus milk:

(a) Milk pre-treated with *Lactobacillus acidophilus*
(b) Used to manage lactose intolerance
(c) Bacterial lactase inactivates lactose into organic acids
(d) Product is sweet yet lactose-free

43. The lowest intestinal absorption rate is of:

(a) Fructose
(b) Glucose
(c) Galactose
(d) All nonsaccharides are absorbed at equal rates

44. Sodium glucose transporter-1 (SGLT-1) is:

(a) Present in apical membrane of enterocytes
(b) Inhibited by ouabain
(c) Inhibited by phlorhizin
(d) All the above

45. In the small intestine, pentoses are absorbed by:

(a) SGLT-1 (b) GLUT-2
(c) ABC protein (d) Simple diffusion

46. The following is true for SGLT-1:

(a) Glucose or galactose can be substrates
(b) Separate binding sites for sodium and monosaccharides
(c) Deficiency leads to glucose-galactose malabsorption
(d) All the above

47. The following is true for fructose absorption in the small intestine:

(a) Via GLUT-5 in the apical membrane
(b) Slower process than glucose and galactose absorption
(c) Energy-independent
(d) All the above

48. GLUT-1 is present in:

(a) Adipose tissue (b) RBC
(c) Skeletal muscle (d) Small intestine

49. The following is not true for GLUT-2:

(a) Present in liver
(b) Insulin-dependent
(c) Sodium-independent
(d) Present in small intestine

50. GLUT-4 is present in:

(a) Skeletal muscle
(b) Cardiac muscle
(c) Adipose tissue
(d) All the above

51. The following is not true for glucose uptake in the brain:

(a) Via GLUT-3
(b) Insulin-dependent
(c) Mainly used in glycolysis
(d) Also used in HMP shunt

52. Hypoxia-inducible transcription factor-1 (HIF-1) enhances GLUT expression in:

(a) Heart
(b) Liver
(c) Placenta
(d) Tumor cells

53. Anaerobic glycolysis occurs in:

(a) Exercising skeletal muscle
(b) RBC
(c) Both
(d) None of the above

54. The following is not true for hexokinase:

(a) Present in extrahepatic tissues
(b) Its isoenzyme glucokinase is present in the liver
(c) Allosterically activated by glucose-6-phosphate
(d) Has a lower K_m for glucose compared with glucokinase

55. The following is not true for phosphofructokinase-1:

(a) ATP is a cosubstrate
(b) ATP is an allosteric activator
(c) Catalyzes an irreversible reaction
(d) Cytosolic

56. The following is true for a positron emission tomography (PET) scan:

(a) Exploits the hexokinase catalyzed reaction
(b) Uses 2-[^{18}F]fluoro-deoxyglucose (FDG)

(c) Detects tumors with enhanced rate of glycolysis
(d) All the above

57. Positive allosteric effectors of phosphofructokinase-1 are:

(a) AMP
(b) Fructose 2,6-bisphosphate
(c) Both
(d) None of the above

58. Glycerol enters glycolysis via:

(a) Glucose-6-phosphate
(b) Fructose-6-phosphate
(c) Dihydroxyacetone phosphate
(d) Pyruvate

59. The number of ATP produced when glycolysis proceeds via Rapoport-Luebering cycle is/are:

(a) Zero
(b) 1
(c) 2
(d) 3

60. The following is a bifunctional enzyme:

(a) Phosphofructokinase-1
(b) Enolase
(c) Phosphoglycerate kinase
(d) Bisphosphoglycerate mutase

61. The following is true for arsenic acid poisoning and glycolysis:

(a) 1-Arseno-3-phosphoglycerate is formed instead of 1,3-bisphosphoglycerate
(b) Glycolysis continues
(c) No ATP is produced
(d) All the above

62. Glycolysis stops in the presence of:

(a) Iodoacetate
(b) Arsenic acid
(c) Both
(d) None of the above

63. The following is not true for glyceraldehyde-3-phosphate dehydrogenase:

(a) Uses NAD^+ hence glycolysis may continue under anaerobic conditions

(b) Contains -SH group
(c) Stimulated by iodoacetate
(d) Inhibited by heavy metals

64. The following is true for fluoride:
(a) Competes with 2-phosphoglycerate for binding to enolase
(b) Added during blood sample collection for glucose estimation
(c) Inhibits glycolysis in bacteria in the oral cavity
(d) All the above

65. The following is not true for neuron specific enolase:
(a) ββ-Isoenzyme
(b) Increased CSF levels in stroke
(c) Increased CSF levels in some CNS tumors
(d) Increased blood levels after head injury

66. The following is not true for pyruvate kinase:
(a) Deficiency leads to hemolytic anemia
(b) Stimulated by alanine
(c) Induced by insulin
(d) Catalyzes a substrate-level phosphorylation

67. Pyruvate dehydrogenase does not require:
(a) TPP
(b) Mg^{2+}
(c) FMN
(d) Lipoic acid

68. Pyruvate dehydrogenase is inhibited by:
(a) Increased $NADH/NAD^+$ ratio
(b) Thiamine deficiency
(c) Arsenite
(d) All the above

69. In anaerobic glycolysis, net ATP yield per molecule of glucose is:
(a) 2
(b) 4
(c) 6
(d) 8

70. In aerobic glycolysis, net ATP yield per molecule of glucose is:

(a) 2 (b) 4
(c) 6 (d) 8

71. The following enzyme requires Fe^{2+}:

(a) Hexokinase (b) Pyruvate dehydrogenase
(c) Aconitase (d) Lactate dehydrogenase

72. Fluoroacetate inhibits:

(a) Hexokinase (b) Pyruvate dehydrogenase
(c) Aconitase (d) Lactate dehydrogenase

73. The following is true for arsenite poisoning:

(a) Inhibits pyruvate dehydrogenase
(b) Inhibits α-ketoglutarate dehydrogenase
(c) Increased hair arsenic
(d) All the above

74. Severe neurological impairment may be seen in deficiency of:

(a) Fumarase
(b) Carbamoyl phosphate synthase-I
(c) Phenylalanine hydroxylase
(d) All the above

75. In each round of the TCA cycle, net ATP yield per molecule of acetyl CoA is:

(a) 2 (b) 10
(c) 12 (d) 14

76. Net ATP yield per molecule of glucose following complete oxidation is:

(a) 8 (b) 18
(c) 28 (d) 38

77. The following B-complex vitamin is not required in the TCA cycle:

(a) Thiamin (Vitamin B_1) (b) Riboflavin (Vitamin B_2)
(c) Niacin (Vitamin B_3) (d) Pyridoxine (Vitamin B_6)

78. The following is an amphibolic pathway:

(a) TCA cycle
(b) Ketogenesis
(c) Catabolism of branched chain amino acids
(d) Glycolysis

79. The following is an inhibitor of the TCA cycle:

(a) Increased NADH/NAD^+ ratio
(b) Increased ATP/ADP ratio
(c) Increased succinyl CoA level
(d) All the above

80. The following is not a source of gluconeogenesis:

(a) Leucine (b) Propionate
(c) Lactate (d) Glycerol

81. Gluconeogenesis occurs:

(a) Mainly in liver, partly in kidneys, not in muscles
(b) Mainly in kidneys, partly in liver, not in muscles
(c) Mainly in muscles, partly in liver, not in kidneys
(d) Mainly in muscles, partly in kidneys, not in liver

82. Gluconeogenesis occurs in:

(a) Cytosol and mitochondria
(b) Cytosol only
(c) Mitochondria only
(d) None of the above

83. The following coenzyme is required by phosphoenolpyruvate carboxykinase:

(a) ATP (b) CTP
(c) UTP (d) GTP

84. Glucose-6-phosphatase is absent in:

(a) Liver and muscle
(b) Adipose tissue and muscle
(c) Liver and adipose tissue
(d) Liver, muscle and adipose tissue

85. The most potent gluconeogenic substrate is:

(a) Aspartate (b) Asparagine
(c) Alanine (d) Proline

86. Cori cycle involves:

(a) Glucose and alanine (b) Malate and aspartate
(c) Glucose and lactate (d) Q cycle

87. For the synthesis of succinyl CoA from propionyl CoA, the following is/are required:

(a) Biotin and Vitamin B_{12} (b) Vitamin B_{12} only
(c) Biotin only (d) None of the above

88. The following metabolic pathways are reciprocally regulated:

(a) Glycolysis and glycogenolysis
(b) Glycolysis and glycogenesis
(c) Glycolysis and gluconeogenesis
(d) Glycolysis and glucuronic acid pathway

89. Glycogen synthase links glucose to the glycogen primer at the:

(a) Non-reducing end
(b) Reducing end
(c) Both ends simultaneously
(d) Both ends alternately

90. Muscle glycogen synthase 'a' is:

(a) Dephosphorylated, active
(b) Dephosphorylated, inactive
(c) Either of the above
(d) None of the above

91. Hepatic glycogenolysis is stimulated by:

(a) Insulin (b) Glucagon and epinephrine
(c) Epinephrine only (d) Glucagon only

92. Glycogenolysis ends in glucose in:

(a) Kidneys (b) Liver
(c) Kidneys and liver (d) None of the above

93. Glycogenolysis ends in glucose-6-phosphate in:
(a) Muscles
(b) Liver
(c) Kidneys
(d) All the above

94. Glycogen phosphorylase requires vitamin:
(a) Thiamin (Vitamin B_1)
(b) Riboflavin (Vitamin B_2)
(c) Pantothenic acid (Vitamin B_5)
(d) Pyridoxine (Vitamin B_6)

95. The following is an allosteric activator of glycogen phosphorylase:
(a) ATP
(b) AMP
(c) Glucose
(d) Glucose-6-phosphate

96. Glycogen phosphorylase 'a' is:
(a) Dephosphorylated, active
(b) Dephosphorylated, inactive
(c) Either of the above
(d) None of the above

97. Glycogen phosphorylase kinase 'a' is:
(a) Dephosphorylated, active
(b) Dephosphorylated, inactive
(c) Either of the above
(d) None of the above

98. The following is a bifunctional enzyme:
(a) Glycogen synthase
(b) Glycogen phosphorylase
(c) Glycogen phosphorylase kinase
(d) Debranching enzyme

99. Under the effect of insulin in muscles, cAMP concentration:
(a) Increases
(b) Decreases
(c) Does not change
(d) May increase or decrease

100. Under the effect of epinephrine in muscles, cAMP concentration:
(a) Increases
(b) Decreases
(c) Does not change
(d) May increase or decrease

101. The δ-subunit of glycogen phosphorylase kinase is:

(a) Ca^{2+}
(b) Calmodulin
(c) cAMP
(d) cAMP-dependent protein kinase

102. Glycogen concentration (w/w) is highest in:

(a) Muscles (b) Liver
(c) Kidneys (d) Adipose tissue

103. Net glycogen content (grams) is highest in:

(a) Muscles (b) Liver
(c) Kidneys (d) Adipose tissue

104. Ketotic hypoglycemia occurs in deficiency of:

(a) Hepatic glycogen synthase
(b) Hepatic glucose-6-phosphatase
(c) Both
(d) None of the above

105. Anderson's disease is due to deficiency of:

(a) Lysosomal acid maltase (b) Debranching enzyme
(c) Branching enzyme (d) Muscle phosphorylase

106. McArdle's syndrome is due to deficiency of:

(a) Lysosomal acid maltase (b) Debranching enzyme
(c) Branching enzyme (d) Muscle phosphorylase

107. Pompe's disease is due to deficiency of:

(a) Lysosomal acid maltase (b) Debranching enzyme
(c) Branching enzyme (d) Muscle phosphorylase

108. Forbe's disease is due to deficiency of:

(a) Lysosomal acid maltase (b) Debranching enzyme
(c) Branching enzyme (d) Muscle phosphorylase

109. In HMP shunt, reducing equivalents are accepted by:

(a) $NADP^+$ (b) NADPH
(c) NAD^+ (d) NADH

110. **HMP shunt generates:**

(a) $NADP^+$
(b) NADPH
(c) NAD^+
(d) NADH

111. **The following is true for HMP shunt:**

(a) Occurs in cytosol, inhibited by insulin
(b) Occurs in mitochondria, inhibited by insulin
(c) Occurs in cytosol, stimulated by insulin
(d) Occurs in mitochondria, stimulated by insulin

112. **HMP shunt operates in:**

(a) Liver
(b) RBC
(c) Lactating mammary glands
(d) All the above

113. **Glucose-6-phosphate dehydrogenase (G6PD) is induced by:**

(a) Glucagon
(b) Epinephrine
(c) Insulin
(d) Cortisol

114. **The first pentose to be synthesized in HMP shunt is:**

(a) Ribose-5-phosphate
(b) Ribulose-5-phosphate
(c) Xylose-5-phosphate
(d) Xylulose-5-phosphate

115. **The following drug(s) should be used with caution in G6PD deficiency:**

(a) Acetanilide
(b) Sulphamethoxazole
(c) Primaquine
(d) All the above

116. **Transketolase requires:**

(a) Biocytin
(b) FAD
(c) $NADP^+$
(d) TPP

117. **Uronic acid pathway generates:**

(a) ATP
(b) NADPH
(c) Both
(d) None of the above

118. **HMP shunt generates:**

(a) Pentoses
(b) NADPH
(c) CO_2
(d) All the above

119. Physiological jaundice of the newborn is due to:
(a) Decreased activity of glucuronyl transferase
(b) Increased activity of glucuronyl transferase
(c) G6PD deficiency
(d) Lack of β-glucuronidase

120. Humans normally lack:
(a) Glucuronyl transferase
(b) L-gulonate dehydrogenase
(c) L-gulonolactone oxidase
(d) 3-Keto-L-gulonate decarboxylase

121. L-Xylulose reductase deficiency leads to:
(a) Infantile scurvy
(b) Essential pentosuria
(c) Congenital hemolytic jaundice
(d) Congenital cataract

122. Polyol synthesis occurs in:
(a) Seminal vesicles
(b) Peripheral nerves
(c) Lens
(d) All the above

123. The following is not true for diabetic cataract:
(a) Decreased sorbitol synthesis in the lens
(b) Glycosylation of crystallin
(c) Partly responsive to aldose reductase inhibitors
(d) Increased lens osmolarity

124. Highest concentration of fructose is in:
(a) Pancreatic juice
(b) CSF
(c) Seminal fluid
(d) Urine

125. The substrate for aldolase 'B' is:
(a) Fructose-1-phosphate
(b) Fructose-1,6-bisphosphate
(c) Fructose-6-phosphate
(d) Fructose-2,6-bisphosphate

126. Hereditary fructose intolerance is due to the deficiency of:
(a) Aldolase 'A'
(b) Aldolase 'B'
(c) Fructokinase
(d) Phosphofructokinase-1

127. Essential fructosuria is due to the deficiency of:

(a) Aldolase 'A'
(b) Aldolase 'B'
(c) Fructokinase
(d) Phosphofructokinase-1

128. The major fate of dietary galactose following intestinal absorption is:

(a) Excretion in urine
(b) Hepatic conversion to glucose
(c) Hepatic conversion to lactose
(d) Hepatic VLDL synthesis

129. The following is derived from UDP-galactose:

(a) UDP-glucose
(b) Galactosyl sphingosine
(c) Lactose
(d) All the above

130. The following is not true for congenital galactosemia:

(a) Deficiency of galactose-1-phosphate uridyl transferase
(b) Absence of reducing sugars in urine
(c) Increased lens galactitol
(d) Decreased cerebral ganglioside synthesis

131. The following are the substrates for lactose synthase:

(a) Glucose and galactose
(b) Glucose and UDP-galactose
(c) UDP-glucose and galactose
(d) UDP-glucose and UDP-galactose

132. Insulin induces the following except:

(a) Fructose-1,6-bisphosphatase
(b) ATP-citrate lyase
(c) G6PD
(d) 6-Phosphogluconate dehydrogenase

133. Insulin promotes the following except:

(a) Glycolysis
(b) Amino acid uptake
(c) Lipogenesis
(d) Ketogenesis

134. Amylin is co-secreted with:

(a) Insulin

(b) Glucagon
(c) Somatostatin
(d) Insulin-like growth factor-1 (IGF-1)

135. Glucose enters pancreatic β-cells via:

(a) Simple diffusion
(b) Active transport
(c) GLUT-2
(d) GLUT-4

136. The following is called a 'glucose sensor' in pancreatic β-cells:

(a) Glucokinase
(b) Insulin
(c) Amylin
(d) Glucose tolerance factor

137. Sulphonylureas act on:

(a) ATP-independent K^+-channels
(b) ATP-dependent K^+-channels
(c) ATP-independent Na^+-channels
(d) ATP-dependent Na^+-channels

138. The release of insulin from pancreatic β-cells requires:

(a) Ca^{2+}
(b) ATP
(c) Microfilaments
(d) All the above

139. The following is partly responsible for the catabolism and inactivation of insulin:

(a) Glucagon
(b) Epinephrine
(c) Thyroxine
(d) Cortisol

140. In the fed state, plasma insulin/glucagon ratio is nearly:

(a) 0.5
(b) 2.0
(c) 30.0
(d) 300.0

141. The following is not true for renal glycosuria:

(a) Inherited defect in proximal tubular SGLT-1
(b) Euglycemia
(c) Osmotic diuresis and glucosuria
(d) Rapidly fatal

142. The following is not true for obese type 2 diabetes mellitus:
(a) Decreased TNF-α in adipocytes
(b) Insulin receptor down-regulation
(c) Hypertriglyceridemia
(d) Risk of non-ketotic hyperosmolar coma

143. Incretin/anti-incretin imbalance occurs in:
(a) Addison's disease
(b) Type 2 diabetes mellitus
(c) Diabetes insipidus
(d) Pheochromocytoma

144. Glycosylated hemoglobin levels in blood are an index of blood glucose control over the past:
(a) 3 hours
(b) 3 days
(c) 3 months
(d) 3 years

145. The following is preferred for assessing pancreatic function in diabetic patients receiving insulin:
(a) Plasma insulin
(b) Plasma C-peptide
(c) Plasma glucose
(d) Plasma glucagon

146. In a diabetic patient who is also suffering from phenylketonuria, the following should be avoided:
(a) Aspartame
(b) Saccharin
(c) Sucralose
(d) All the above

147. The following sugar (linked to several fatty acid molecules) is commonly used to synthesize artificial fat:
(a) Glucose
(b) Lactose
(c) Maltose
(d) Sucrose

148. The major source of plasma free fatty acids is:
(a) Liver
(b) Adipose tissue
(c) Muscle
(d) Brain

149. The omega system of nomenclature of a fatty acid begins with the:
(a) Methyl-carbon atom
(b) Carboxyl-carbon atom
(c) Carbon atom next to the methyl-carbon atom
(d) Carbon atom next to the carboxyl-carbon atom

150. Butter and peanut oil are rich in the following saturated fatty acids:

(a) Arachidic acid and butyric acid, respectively
(b) Butyric acid and arachidic acid, respectively
(c) Arachidonic acid and butyric acid, respectively
(d) Butyric acid and arachidonic acid, respectively

151. The following are associated with an increased risk of cardiovascular disease:

(a) Diet rich in saturated fatty acids
(b) Diet rich in *trans*-fatty acids
(c) Increased plasma oxidized-LDL
(d) All the above

152. Phrenoderma results from a dietary deficiency of:

(a) Essential amino acids (b) Essential fatty acids
(c) Phytanic acid (d) Propionic acid

153. On a balanced diet, the following fatty acid is not truly essential:

(a) Arachidonic acid (b) Linoleic acid
(c) Linolenic acid (d) Eicosapentenoic acid

154. In polyunsaturated fatty acids (PUFA), double bonds generally occur after every:

(a) Two carbon atoms (b) Three carbon atoms
(c) Four carbon atoms (d) Six carbon atoms

155. In patients with increased plasma triglycerides, the recommended diet should be rich in:

(a) ω-6 Fatty acids (b) ω-3 Fatty acids
(c) Saturated fatty acid (d) *Trans*-fatty acids

156. In patients with increased plasma total cholesterol, the recommended diet should be rich in:

(a) ω-6 Fatty acids (b) ω-3 Fatty acids
(c) Saturated fatty acids (d) *Trans*-fatty acids

157. Lipuria may occur in:

(a) Fracture of long bones
(b) Phosphorus poisoning
(c) Lipoid nephrosis
(d) All the above

158. The following is not true for pancreatic lipase:

(a) Also called steapsin
(b) An α-lipase
(c) Preferred substrates are short chain fatty acids
(d) Requires a co-lipase

159. Pancreatic lipase hydrolyzes triacylglycerol preferentially at:

(a) C1 and C2
(b) C2 and C3
(c) C1 and C3
(d) C1, C2 and C3

160. The following is most commonly observed in steatorrhea:

(a) Increased bleeding tendency
(b) Night blindness
(c) Bone pain
(d) Painful micturition

161. The following is true for intestinal absorption of short chain fatty acids:

(a) Not absorbed
(b) Absorbed mainly via portal circulation
(c) Absorbed mainly via lacteals as chylomicrons
(d) Absorbed after getting converted to long chain fatty acids

162. Urine may contain chylomicrons in:

(a) Lymphatic filariasis
(b) Falciparum malaria
(c) Dengue fever
(d) Intestinal tuberculosis

163. The following is true for intestinal absorption of long chain fatty acids:

(a) Not absorbed
(b) Absorbed mainly via portal circulation
(c) Absorbed mainly via lacteals as chylomicrons
(d) Absorbed after getting converted to short chain fatty acids

164. The following vitamin is not involved in β-oxidation of fatty acids:

(a) Pantothenic acid (Vitamin B_5)
(b) Riboflavin (Vitamin B_2)
(c) Niacin (Vitamin B_3)
(d) Pyridoxine (Vitamin B_6)

165. The following is not true for thiokinases:

(a) Also called acyl CoA synthetases
(b) Found in mitochondria only
(c) Catalyze fatty acid activation
(d) Specificity differs with fatty acid chain length

166. Carnitine synthesis requires:

(a) Lysine only
(b) Methionine only
(c) Lysine and methionine
(d) None of the above

167. The following is true for patients suffering from Reye's syndrome:

(a) Defect in plasma membrane carnitine transporter
(b) Increased urinary acyl carnitine
(c) Mutation in carnitine palmitoyl transferase-II gene
(d) None of the above

168. The following deficiency occurs in patients suffering from Reye's syndrome:

(a) Acyl CoA dehydrogenase
(b) Enoyl CoA hydratase
(c) β-Hydroxyacyl CoA dehydrogenase
(d) β-Keto thiolase

169. The commonest deficiency in Reye's syndrome is of:

(a) Short chain acyl CoA dehydrogenase
(b) Medium chain acyl CoA dehydrogenase
(c) Long chain acyl CoA dehydrogenase
(d) None of the above

170. Trimetazidine inhibits:

(a) Acyl CoA dehydrogenase
(b) Enoyl CoA hydratase
(c) β-Hydroxyacyl CoA dehydrogenase
(d) β-Keto thiolase

171. Net ATP yield is 129 with complete oxidation of:

(a) $C_{15}H_{31}COOH$
(b) $C_{17}H_{35}COOH$
(c) $C_{19}H_{39}COOH$
(d) $C_{13}H_{27}COOH$

172. In peroxisomal fatty acid oxidation, chain shortening occurs upto:

(a) 12C
(b) 10C
(c) 8C
(d) 6C

173. The following is true for peroxisomal fatty acid oxidation:

(a) Produces H_2O_2
(b) Produces octanoyl CoA
(c) Specific for long chain fatty acids
(d) All the above

174. The following is a product of α-oxidation of fatty acids:

(a) CO_2
(b) ATP
(c) Both
(d) None of the above

175. The following is true for Refsum's disease:

(a) Accumulation of phytanic acid
(b) Deficiency of α-hydroxylating monooxygenase
(c) Neurological disturbance
(d) All the above

176. α-Oxidation of fatty acid requires:

(a) Vitamin D
(b) Thiamin (Vitamin B_1)
(c) Vitamin C
(d) Biotin

177. a-Oxidation of fatty acid requires:

(a) Fe^{2+}
(b) Zn^{2+}
(c) Ca^{2+}
(d) All the above

178. α-Oxidation of fatty acid occurs in:

(a) Cytosol
(b) Mitochondria
(c) Endoplasmic reticulum
(d) Extracellular matrix

179. ω-Oxidation of fatty acid requires:

(a) NADPH
(b) CO_2
(c) Cytochrome P450
(d) All the above

180. ω-Oxidation of fatty acid occurs in:

(a) Endoplasmic reticulum
(b) Mitochondria
(c) Cytosol
(d) Extracellular matrix

181. Oxidation of fatty acids with odd number of carbon atoms occurs in:

(a) Endoplasmic reticulum
(b) Mitochondria
(c) Cytosol
(d) Extracellular matrix

182. Oxidation of unsaturated fatty acids occurs via:

(a) α-Oxidation
(b) β-Oxidation
(c) ω-Oxidation
(d) Any of the above

183. Acyl CoA carboxylase requires:

(a) Biotin
(b) Pyrixodine
(c) Pantothenic acid
(d) Vitamin C

184. The following is not true for acyl CoA carboxylase:

(a) Catalyzes the committed step in fatty acid synthesis
(b) Glucagon promotes enzyme phosphorylation
(c) Active in phosphorylated state
(d) Requires ATP and biotin

185. A 'metabolon' is:

(a) A metabolic disorder
(b) A metabolic pathway with multiple regulatory steps
(c) A multienzyme complex
(d) Metabolic integration between various organs

186. NADPH is required for fatty acid:

(a) Chain elongation
(b) Desaturation
(c) Hydroxylation
(d) All the above

187. Acyl carrier protein contains:

(a) Pantothenic acid
(b) Lipoic acid
(c) Arachidonic acid
(d) Palmitic acid

188. Fatty acid elongation occurs in:

(a) Endoplasmic reticulum
(b) Mitochondria
(c) Both
(d) None of the above

189. By chain elongation, C24 fatty acids can be synthesized in:

(a) Liver
(b) Kidneys
(c) Adipose tissue
(d) Brain

190. Stearoyl CoA desaturase is induced by:

(a) Insulin
(b) Hydrocortisone
(c) Triiodothyronine
(d) All the above

191. During fatty acid desaturation in humans, the first double bond is introduced between:

(a) C9-C10
(b) C8-C9
(c) C4-C5
(d) C2-C3

192. Cytochrome b_5 is required during fatty acid:

(a) Chain elongation
(b) Chain desaturation
(c) Hydroxylation
(d) Oxidation

193. Compared with white adipose tissue, brown adipose tissue:

(a) Is more vascular
(b) Generates more heat
(c) Has higher metabolic activity
(d) All the above

194. Triacylglycerol stored in human adipose tissue generally has:

(a) Palmitic acid at C1 but oleic acid at C2 and C3
(b) Oleic acid at C1 but palmitic acid at C2 and C3
(c) Palmitic acid at C1, C2 and C3
(d) Oleic acid at C1, C2 and C3

195. Lipoprotein lipase is located in:

(a) Cytosol of adipocytes
(b) Cell membrane of adipocytes

(c) Endothelial cell surface
(d) Plasma lipoproteins

196. Hormone sensitive lipase is located in:

(a) Cytosol of adipocytes
(b) Cell membrane of adipocytes
(c) Endothelial cell surface
(d) Plasma lipoproteins

197. Hormone sensitive lipase is stimulated by the following except:

(a) ACTH (b) Insulin
(c) Glucagon (d) Epinephrine

198. Quantitatively, the most important way of cholesterol excretion is in the form of:

(a) Fecal sterols (b) Shedding by skin
(c) Bile acids (d) Urinary steroids

199. Plasma cholesterol is:

(a) 100% unesterified
(b) 100% esterified
(c) 30% unesterified, 70% esterified
(d) 30% esterified, 70% unesterified

200. In biomembranes, cholesterol is:

(a) 100% unesterified
(b) 100% esterified
(c) 30% unesterified, 70% esterified
(d) 30% esterified, 70% unesterified

201. Plasma LDL-cholesterol is mainly esterified with:

(a) Linoleate (b) Linolenate
(c) Arachidonate (d) Palmitate and palmitoleate

202. Cytosolic cholesterol is mainly esterified with:

(a) Linoleate (b) Linolenate
(c) Arachidonate (d) Palmitate and palmitoleate

203. Cholesterol is a precursor of vitamin:

(a) A (b) D
(c) E (d) K

204. Cholesterol is synthesized in:

(a) Gonads (b) Skin
(c) Adrenal cortex (d) All the above

205. HMG CoA synthase is located in:

(a) Cytosol (b) Mitochondria
(c) Both (d) None of the above

206. HMG CoA can be obtained by the catabolism of:

(a) Lysine (b) Leucine
(c) Isoleucine (d) Valine

207. The parent ketone body is:

(a) Acetoacetate (b) β-Hydroxy butyrate
(c) Acetone (d) Dihydroxyacetone

208. Ketogenesis occurs in hepatocyte:

(a) Cytosol (b) Endoplasmic reticulum
(c) Mitochondria (d) Golgi complex

209. The normal plasma concentration of ketone bodies is:

(a) 1-3 mg/dl (b) 3-6 mg/dl
(c) 6-9 mg/dl (d) 9-12 mg/dl

210. The following has a characteristic fruity odour:

(a) Acetoacetate (b) β-Hydroxy butyrate
(c) Acetone (d) Dihydroxyacetone

211. Increased circulating free fatty acids are observed in:

(a) Diabetic ketoacidosis
(b) Starvation
(c) High-fat low-carbohydrate diet
(d) All the above

212. The following cannot use ketone bodies as a source of energy:

(a) Liver (b) Kidneys
(c) Muscles (d) Brain

213. The following does not possess thiophorase:

(a) Liver (b) Kidneys
(c) Muscles (d) Brain

214. The following is a pseudoketone:

(a) Acetoacetate (b) β-Hydroxy butyrate
(c) Acetone (d) Dihydroxyacetone

215. In severe diabetic ketoacidosis, insulin is administered by the following route:

(a) Subcutaneous (b) Intramuscular
(c) Intravenous infusion (d) Pulmonary

216. The following can synthesize phospholipids except:

(a) Neurons (b) Mature RBC
(c) Endothelial cells (d) Pancreatic β-cells

217. Phosphatidylethanolamine N-methyl transferase is present in:

(a) Cytosol (b) Mitochondria
(c) Endoplasmic reticulum (d) Lysosomes

218. The normal lecithin/sphingomyelin (L/S) ratio in amniotic fluid at 34 weeks of gestation is:

(a) 0.5 (b) 1.0
(c) 1.5 (d) 2.0

219. The following triggers the release of Ca^{2+} from the endoplasmic reticulum:

(a) Inositol triphosphate (b) 1,2-Diacylglycerol
(c) Protein kinase C (d) Calmodulin

220. Protein kinase C is activated by:

(a) Inositol triphosphate
(b) 1,2-Diacylglycerol
(c) Phosphatidylinositol bisphosphate
(d) Calmodulin

221. Cardiolipin is mainly localized in:

(a) Outer mitochondrial membrane
(b) Inner mitochondrial membrane

(c) Mitochondrial matrix
(d) Mitochondrial intermembrane space

222. The following is not synthesized in Zellweger disease:
(a) Cholesterol
(b) Plasmalogens
(c) Glycogen
(d) Pyrimidines

223. The content of ethanolamine plasmalogen is more than that of choline plasmalogen in:
(a) Heart muscle
(b) Myelin
(c) Both
(d) None of the above

224. The content of choline plasmalogen is more than that of ethanolamine plasmalogen in:
(a) Heart muscle
(b) Myelin
(c) Both
(d) None of the above

225. Phosphorus is absent in:
(a) RNA
(b) Lecithin
(c) Sphingomyelin
(d) Ceramide

226. The fatty acid in ceramide is:
(a) Palmitic acid
(b) Myristic acid
(c) Behenic acid
(d) Arachidonic acid

227. The fatty acid in ceramide is:
(a) C16
(b) C14
(c) C22
(d) C20

228. Sphingomyelin of grey matter mainly contains:
(a) Lignoceric acid
(b) Nervonic acid
(c) Stearic acid
(d) Linoleic acid

229. The following are neutral sphingolipids except:
(a) Cerebrosides
(b) Globosides
(c) Gangliosides
(d) None of the above

230. The following are acidic sphingolipids except:
(a) Cerebrosides
(b) Gangliosides
(c) Sulphatides
(d) None of the above

231. The enzyme deficient in Niemann-Pick's disease is:

(a) Sphingomyelinase (b) Galactocerebrosidase
(c) Glucocerebrosidase (d) α-Galactosidase

232. The enzyme deficient in Krabbe's disease (Globoid leukodystrophy) is:

(a) Sphingomyelinase (b) Galactocerebrosidase
(c) Glucocerebrosidase (d) α-Galactosidase

233. The enzyme deficient in Gaucher's disease is:

(a) Sphingomyelinase (b) Galactocerebrosidase
(c) Glucocerebrosidase (d) β-Galactosidase

234. The enzyme deficient in Fabry's disease is:

(a) Sphingomyelinase (b) Galactocerebrosidase
(c) Glucocerebrosidase (d) α-Galactosidase

235. The enzyme deficient in metachromatic leukodystrophy is:

(a) Sulphatidase (b) Hexosaminidase A
(c) β-Galactosidase (d) Sphingomyelinase

236. The enzyme deficient in G_{M1} gangliosidosis is:

(a) Sulphatidase (b) Hexosaminidase A
(c) β-Galactosidase (d) Sphingomyelinase

237. G_{M2} gangliosidosis is:

(a) Gaucher's disease
(b) Tay-Sach's disease
(c) Niemann-Pick's disease
(d) Metachromatic leukodystrophy

238. Recombinant glucocerebrosidase is used to treat:

(a) Gaucher's disease
(b) Tay-Sach's disease
(c) Niemann-Pick's disease
(d) Metachromatic leukodystrophy

239. Chylomicrons transport triacylglycerol from:

(a) Liver to small intestine only
(b) Small intestine to liver only

(c) Small intestine to extrahepatic tissues
(d) Liver to extrahepatic tissues

240. Very low density lipoproteins (VLDL) transport triacylglycerol from:
(a) Liver to small intestine only
(b) Small intestine to liver
(c) Small intestine to extrahepatic tissues
(d) Liver to extrahepatic tissues

241. Apo B-48 is present in:
(a) Chylomicrons (b) VLDL
(c) LDL (d) HDL

242. Nascent chylomicrons receive apo E and apo C from:
(a) VLDL (b) LDL
(c) IDL (d) HDL

243. The major lipid in chylomicrons is:
(a) Dietary triglycerides (b) Hepatic triglycerides
(c) Cholesterol (d) Cholesterol esters

244. The major lipid in VLDL is:
(a) Dietary triglycerides (b) Hepatic triglycerides
(c) Cholesterol (d) Cholesterol esters

245. Nascent VLDL does not contain:
(a) Triglycerides (b) Apo B-100
(c) Apo A-1 (d) Apo C-II

246. Lipoprotein lipase acts on:
(a) Chylomicrons (b) VLDL
(c) Both (d) None of the above

247. Cholesterol ester transfer protein (CETP) transfers cholesterol esters from:
(a) HDL to VLDL (b) VLDL to HDL
(c) Chylomicrons to HDL (d) HDL to chylomicrons

248. The following is true for LDL type B:

(a) Small and dense
(b) Prone to oxidation
(c) Highly atherogenic
(d) All the above

249. LDL receptors recognize:

(a) Apo B-48
(b) Apo C-I
(c) Apo C-II
(d) Apo B-100

250. LDL is cleared from the circulation as follows:

(a) 50% by LDL receptors and 50% by scavenger receptors of phagocytes
(b) 75% by LDL receptors and 25% by scavenger receptors of phagocytes
(c) 25% by LDL receptors and 75% by scavenger receptors of phagocytes
(d) None of the above

251. Lecithin cholesterol acyl transferase (LCAT) is activated by:

(a) Apo A-I
(b) Apo C-I
(c) Apo B-100
(d) Apo B-48

252. Acyl cholesterol acyl transferase (ACAT) catalyzes the esterification of cholesterol with:

(a) Monounsaturated fatty acids
(b) Polyunsaturated fatty acids
(c) Saturated fatty acids
(d) Any of the above

253. Acyl cholesterol acyl transferase (ACAT) is located:

(a) In the hepatocyte cytosol
(b) In plasma
(c) On the endothelial cell surface of extrahepatic tissues
(d) On the endothelial cell surface of hepatic capillaries

254. Highest concentration of phospholipids are found in:

(a) Chylomicrons
(b) HDL
(c) VLDL
(d) LDL

255. Nascent HDL does not contain:

(a) Apo C
(b) Apo A
(c) Apo E
(d) Free cholesterol

256. Lipoprotein lipase is activated by:

(a) Apo B-48
(b) Apo C-I
(c) Apo C-II
(d) Apo B-100

257. Hepatic HDL-uptake occurs via:

(a) LDL receptors
(b) SR-B1 receptors
(c) Macrophage scavenger receptors
(d) Non-receptor mediated endocytosis

258. In reverse cholesterol transport, the route of cholesterol flow is from:

(a) HDL_3 to HDL_2 to liver
(b) HDL_2 to HDL_3 to liver
(c) Liver to HDL_2 to HDL_3
(d) Liver to HDL_3 to HDL_2

259. The following is not a risk factor for atherosclerosis:

(a) Increased plasma homocysteine
(b) Increased plasma lipoprotein(a)
(c) Increased plasma LDL-cholesterol
(d) Increased plasma HDL-cholesterol

260. The following inhibits intestinal cholesterol absorption:

(a) Atorvastatin
(b) Nicotinic acid
(c) Cholestyramine
(d) Ezetimibe

261. The following inhibits HMG CoA reductase:

(a) Atorvastatin
(b) Nicotinic acid
(c) Cholestyramine
(d) Ezetimibe

262. The following is a bile acid sequestering agent:

(a) Atorvastatin
(b) Nicotinic acid
(c) Cholestyramine
(d) Ezetimibe

263. The following increases circulating HDL-cholesterol:

(a) Atorvastatin (b) Nicotinic acid
(c) Cholestyramine (d) Ezetimibe

264. Primary type I hyperlipoproteinemia is characterized by increased plasma:

(a) Chylomicrons (b) LDL
(c) LDL+VLDL (d) VLDL

265. Primary type IIa hyperlipoproteinemia is characterized by increased plasma:

(a) Chylomicrons (b) LDL
(c) LDL+VLDL (d) VLDL

266. Primary type IIb hyperlipoproteinemia is characterized by increased plasma:

(a) Chylomicrons (b) LDL
(c) LDL+VLDL (d) VLDL

267. Primary type III hyperlipoproteinemia is characterized by increased plasma:

(a) Chylomicrons (b) LDL
(c) LDL+VLDL (d) Remnant VLDL

268. Primary type IV hyperlipoproteinemia is characterized by increased plasma:

(a) Chylomicrons (b) LDL
(c) LDL+VLDL (d) VLDL

269. Primary type V hyperlipoproteinemia is characterized by increased plasma:

(a) Chylomicrons (b) LDL
(c) LDL+VLDL (d) VLDL+Chylomicrons

270. 'Broad beta disease' is:

(a) Primary type I hyperlipoproteinemia
(b) Primary type III hyperlipoproteinemia
(c) Primary type IV hyperlipoproteinemia
(d) Primary type V hyperlipoproteinemia

271. Generally, the plasma lipid profile in secondary hyperlipoproteinemias resembles that of primary hyperlipoproteinemia type:

(a) I (b) II
(c) III (d) IV

272. Secondary hyperlipoproteinemia resembling primary hyperlipoproteinemia type IV may occur in:

(a) Diabetes mellitus (b) Chronic alcoholism
(c) von Gierke's disease (d) All the above

273. Secondary hyperlipoproteinemia may occur in the following except:

(a) Acromegaly
(b) Prolonged use of oral contraceptives
(c) Uremia
(d) Hyperthyroidism

274. The following may be observed in primary type I hyperlipoproteinemia:

(a) Lipoprotein lipase deficiency
(b) Apo C-II deficiency
(c) Presence of lipoprotein lipase inhibitor
(d) All the above

275. Secondary hyperlipoproteinemia resembling primary hyperlipoproteinemia type IIa may occur in the following except:

(a) Diabetes mellitus (b) Nephrotic syndrome
(c) Hyperthyroidism (d) Anorexia nervosa

276. Secondary hyperlipoproteinemia resembling primary hyperlipoproteinemia type IIa may occur in the following:

(a) Esophageal carcinoma
(b) Primary hepatocellular carcinoma
(c) Carcinoma prostate
(d) Carcinoma cervix

277. Tangier's disease is characterized by deficiency of circulating:

(a) LDL
(b) VLDL
(c) Chylomicrons
(d) HDL

278. Fatty liver may occur in:

(a) Kwashiorkor
(b) Chronic alcoholism
(c) Diabetes mellitus
(d) All the above

279. The normal hepatic triglyceride content (w/w) is:

(a) 1-5%
(b) 5-10%
(c) 10-15%
(d) 15-20%

280. The following is not true for saposins:

(a) Proteins facilitating the catabolism of glycosphingolipids
(b) Mainly located in cytosol
(c) Saposins A, B, C and D are obtained by cleavage of a single precursor protein
(d) Mutations in saposin gene may result in Gaucher's disease, Tay-Sach's disease and metachromatic leukodystrophy

281. Of the total body weight, proteins constitute:

(a) 25%
(b) 50%
(c) 75%
(d) 98%

282. The following is not true for a peptide bond:

(a) 'Cis' in nature
(b) Partial double bond character
(c) No freedom of rotation
(d) Freedom of rotation of side chains is possible for C and N

283. The bond between g-COOH group of glutamate and a-NH_2 group of cysteine in glutathione is:

(a) A true peptide bond
(b) An isopeptide bond
(c) Not a peptide bond
(d) An unstable bond

284. The following is true for gramicidin:

(a) Circular structure
(b) A protein
(c) An antibiotic
(d) All the above

285. In an insulin molecule, the number of disulphide bonds are:

(a) 1
(b) 2
(c) 3
(d) 4

286. In an insulin molecule, the disulphide bonds are:

(a) 2 Interchain and 1 intrachain
(b) 1 Interchain and 2 intrachain
(c) 3 Interchain
(d) 3 Intrachain

287. The intrachain disulphide bond of insulin connects two cysteine residues at positions:

(a) 5 and 10
(b) 6 and 11
(c) 7 and 12
(d) 8 and 13

288. The interchain disulphide bonds of insulin connect two cysteine residues at positions:

(a) A7 with B7 and A20 with B19
(b) A7 with B7 and A19 with B20
(c) A7 with B19 and A20 with B7
(d) A7 with B19 and A19 with B7

289. The following is true for porcine insulin with the C-terminal alanine removed:

(a) Antigenically similar to human insulin
(b) Does not elicit antibodies after repeated administration in humans
(c) Known as de-alaninated porcine insulin
(d) All the above

290. In a protein, the following is not true for an α-helix:

(a) Commonest secondary structure
(b) Most stable secondary structure
(c) Highest proportion is found in chymotrypsin
(d) Generally right-handed

291. In a protein, the following is not true for an α-helix:

(a) Least in hemoglobin and myoglobin
(b) Each turn has 3.6 amino acids

(c) Stabilized by H-bonds
(d) Proline does not allow α-helix formation

292. The following is not true for carbonic anhydrase:
(a) Zn-containing enzyme
(b) Contains only parallel β-sheets
(c) Inhibited by acetazolamide
(d) Catalyzes a reversible reaction

293. Zinc ions in carbonic anhydrase are coordinated with:
(a) Arginine (b) Lysine
(c) Asparagine (d) Histidine

294. A compact, globular functional unit of a protein is called a:
(a) Secondary structure (b) Motif
(c) Domain (d) Catalytic site

295. The first protein to be sequenced was:
(a) Chymotrypsin (b) Insulin
(c) Globin (d) Albumin

296. The molecular weight of a protein may be determined by:
(a) SDS-PAGE (b) LC-MS
(c) MALDI-TOF (d) Any of the above

297. In a protein, the N-terminal amino acid can be identified by:
(a) Dansyl chloride (b) Fluoro-dinitrobenzene
(c) Any of the above (d) None of the above

298. Sanger's reagent is:
(a) Fluoro-dinitobenzene (b) Dansyl chloride
(c) Cyanogen bromide (d) Phenyl isothiocyanate

299. Edman's reagent is:
(a) Fluoro-dinitobenzene (b) Dansyl chloride
(c) Cyanogen bromide (d) Phenyl isothiocyanate

300. Carboxypeptidase A does not act if the C-terminal amino acid is:
(a) Lysine (b) Proline
(c) Arginine (d) Any of the above

301. Carboxypeptidase B acts only if the penultimate C-terminal amino acid is:

(a) Lysine
(b) Proline
(c) Arginine
(d) Any of the above

302. Cyanogen bromide attacks:

(a) N-side of methionine
(b) N-side of cysteine
(c) C-side of methionine
(d) C-side of cysteine

303. The first major protein synthesized chemically was:

(a) Pepsinogen
(b) Insulin
(c) Growth hormone
(d) Albumin

304. The following protein shows reversible denaturation:

(a) Ribonuclease
(b) Immunoglobulin
(c) Both the above
(d) None of the above

305. The following can result in protein denaturation:

(a) Heat
(b) UV rays
(c) Urea
(d) All the above

306. Pyroglutamic acid is present in:

(a) GABA
(b) GSH
(c) TRH
(d) Prothrombin

307. UV absorption by proteins at 280 nm is due to:

(a) Aromatic amino acids
(b) Basic amino acids
(c) Acidic amino acids
(d) Sulphur-containing amino acids

308. Chaperones perform the following functions except:

(a) Burial of hydrophobic core of proteins
(b) Moving proteins to their cellular destinations
(c) Refolding of partially unfolded mature proteins
(d) Destruction of aged proteins

309. The following proteins are known to be associated with abnormal folding resulting in disease:

(a) CFTR
(b) α_1-AT
(c) β-Amyloid
(d) All the above

310. The normal plasma albumin/globulin (A/G) ratio is:

(a) 1.2 : 1 (b) 2.2 : 1
(c) 1 : 1.2 (d) 1 : 2.2

311. Plasma albumin transports:

(a) Free fatty acids (b) Unconjugated bilirubin
(c) Steroids (d) All the above

312. Reversal of plasma A/G ratio may occur in:

(a) Hepatic cirrhosis
(b) Nephrotic syndrome
(c) Protein energy malnutrition
(d) All the above

313. Hyperglobulinemia may occur in:

(a) Tuberculosis (b) Rheumatoid arthritis
(c) Hepatic cirrhosis (d) All the above

314. The following protein is required for the proximal tubular reabsorption of albumin normally filtered by the glomerulus:

(a) Megalin (b) Cubilin
(c) Both (d) None of the above

315. Pepsin cleaves peptide bonds where amino group is contributed by:

(a) Aromatic amino acids (b) Aspartate
(c) Glutamate (d) Any of the above

316. Pepsinogen A is activated by:

(a) H^+ (b) Pepsin
(c) Both (d) None of the above

317. Pepsinogen A is secreted by gastric:

(a) G cells (b) Parietal cells
(c) Chief cells (d) All the above

318. The following is not true for rennin:

(a) Found in infants
(b) Cleaves casein

(c) Stimulates the synthesis of angiotensin II
(d) An endopeptidase

319. Trypsin cleaves peptide bonds where carboxyl group is contributed by:

(a) Arginine
(b) Lysine
(c) Any the above
(d) None of the above

320. Chymotrypsin cleaves peptide bonds where carboxyl group is contributed by:

(a) Aromatic amino acids
(b) Leucine or methionine
(c) Asparagine or histidine
(d) Any of the above

321. The following is true for pro-carboxypeptidase A:

(a) Secreted by pancreas
(b) Activated by trypsin
(c) Active enzyme is an endopeptidase that cleaves C-terminal peptide bonds where the terminal amino acid is an aromatic or an aliphatic amino acid
(d) All the above

322. The following is true for pro-carboxypeptidase B:

(a) Secreted by pancreas
(b) Activated by trypsin
(c) Active enzyme is an endopeptidase that cleaves C-terminal peptide bonds where the terminal amino acid is arginine or lysine
(d) All the above

323. The intestinal absorption of amino acids requires:

(a) Pyridoxine (Vitamin B_6)
(b) Niacin (Vitamin B_3)
(c) Thiamin (Vitamin B_1)
(d) Folic acid

324. The following is not true for Hartnup's disease:

(a) Neutral aminoaciduria
(b) Pellagra-like features
(c) Genetic defect in epithelial amino acid transport
(d) Inability to absorb basic amino acids and sulphur-containing amino acids

325. The following is true for cystinuria:
- (a) Neutral aminoaciduria
- (b) Pellagra-like features
- (c) Acquired defect in epithelial amino acid transport
- (d) Inability to absorb basic amino acids and sulphur-containing amino acids

326. The following is not true for celiac sprue:
- (a) Sensitivity to gluten protein of cereals
- (b) Increased serum anti-gluten antibodies
- (c) Partly responsive to corticosteroids
- (d) Highly responsive to more frequent administration of gluten-rich food

327. Increased muscle proteolysis is seen in:
- (a) Extensive thermal injury
- (b) Septicemia
- (c) Starvation
- (d) All the above

328. The following is a purely ketogenic amino acid:
- (a) Leucine
- (b) Isoleucine
- (c) Valine
- (d) Proline

329. The following is a glucogenic+ketogenic amino acid:
- (a) Threonine
- (b) Asparagine
- (c) Lysine
- (d) Serine

330. The following is a purely glucogenic amino acid:
- (a) Isoleucine
- (b) Leucine
- (c) Tyrosine
- (d) Methionine

331. The following amino acid is a substrate for transamination:
- (a) Isoleucine
- (b) Proline
- (c) Threonine
- (d) Lysine

332. Transamination requires the following coenzyme:
- (a) Pyridoxal-5-PO_4 (PLP)
- (b) Thiamin pyrophosphate (TPP)
- (c) NAD^+
- (d) $NADP^+$

333. In a patient suffering from acute myocardial infarction, the following will be increased in serum:

(a) AST (SGOT) only (b) ALT (SGPT) only
(c) Both (d) None of the above

334. The following is not true for transamination:

(a) Involves a pair of substrates
(b) Generally, glutamate and α-ketoglutarate are one of the pairs
(c) There is a net loss of an amino group
(d) Involves a Schiff-base intermediate

335. The following enzyme catalyzes oxidative deamination:

(a) Glutamate dehydrogenase
(b) D-amino acid oxidase
(c) L-amino acid oxidase
(d) All the above

336. The prosthetic group in L-amino acid oxidase is:

(a) FMN (b) FAD
(c) NAD^+ (d) $NADP^+$

337. The prosthetic group in D-amino acid oxidase is:

(a) FMN (b) FAD
(c) NAD^+ (d) $NADP^+$

338. The following is not true for glutamate dehydrogenase:

(a) Important for oxidative deamination
(b) Zn^{2+}-containing enzyme
(c) Requires NAD^+
(d) Inhibited by ADP and GDP

339. The following amino acid undergoes non-oxidative deamination:

(a) Serine (b) Cysteine
(c) Histidine (d) All the above

340. Non-oxidative deamination of amino acids requires:

(a) Thiamin pyrophosphate (TPP)
(b) FAD

(c) Pyridoxal-5-PO_4 (PLP)
(d) NAD

341. Urocanic acid is formed during the deamination of:

(a) Cysteine (b) Threonine
(c) Proline (d) Histidine

342. The following is not true for carbamoyl phosphate synthase-I:

(a) Allosterically inhibited by N-acetyl glutamate
(b) A mitochondrial enzyme
(c) Catalyzes the 1st reaction of urea cycle
(d) Catalyzes an ATP-dependent reaction

343. The following is not true for carbamoyl phosphate synthase-II:

(a) Not affected by N-acetyl glutamate
(b) A cytosolic enzyme
(c) Forms carbamoyl phosphate by using amide group of asparagine
(d) Required for pyrimidine biosynthesis

344. Ornithine transcarbamoylase is located in:

(a) Mitochondria (b) Cytosol
(c) Nucleus (d) Lysosomes

345. In each urea molecule, the two nitrogen atoms are derived from:

(a) One from ammonia and the other from aspartate
(b) One from ammonia and the other from arginine
(c) Both from ammonia
(d) Both from glutamine

346. N-acetyl glutamate synthetase requires the following as substrate and activator, respectively:

(a) Glutamate and arginine
(b) Glutamate and asparagine
(c) Glutamine and arginine
(d) Glutamine and asparagine

347. Ornithine-citrulline antiport is present in:

(a) Plasma membrane
(b) Nuclear membrane
(c) Mitochondrial membrane
(d) Lysosomal membrane

348. Ornithine-citrulline transporter moves:

(a) Ornithine from cytosol to mitochondria and citrulline from mitochondria to cytosol
(b) Citrulline from cytosol to mitochondria and ornithine from mitochondria to cytosol
(c) Both ornithine and citrilline from cytosol to mitochondria
(d) Both ornithine and citrilline from mitochondria to cytosol

349. The following is observed in hyperammonemia type II:

(a) Deficiency of hepatic ornithine transcarbamoylase
(b) Increased blood ammonia levels
(c) Increased cerebral GABA levels
(d) All the above

350. Glutaminase catalyzed reaction is most important in:

(a) Brain
(b) Kidneys
(c) Small intestine
(d) Liver

351. Glycine is not a precursor of:

(a) GABA
(b) GSH
(c) Heme
(d) Bile acids

352. The following is true for glycine:

(a) Simplest amino acid
(b) Non-essential amino acid
(c) Glucogenic amino acid
(d) All the above

353. Deficiency of glycine cleavage complex leads to the following except:

(a) Hyperglycinemia
(b) Mental retardation
(c) Ketosis
(d) Death during infancy

354. The following is true for glycine transaminase deficiency:

(a) Increased production of oxalate

(b) Primary hyperoxaluria and risk of renal oxalate stones
(c) Partly responsive to vitamin B_6
(d) All the above

355. The reversible interconversion of serine and glycine requires:
(a) NAD^+
(b) $NADP^+$
(c) Pyridoxal-5-PO_4 (PLP)
(d) Biotin

356. In a purine ring, glycine contributes:
(a) N7, C4 and C5
(b) N3, C4 and C5
(c) N7, C8 and C9
(d) N1, C4 and C5

357. In humans, D-alanine is found:
(a) In proteins
(b) Free in circulation
(c) Colonic bacterial cell wall
(d) All the above

358. β-Alanine is synthesized from:
(a) α-Alanine
(b) Uracil
(c) β-Hydroxybutyric acid
(d) Betaine

359. β-Alanine is present in:
(a) Pantothenic acid
(b) Carnosine
(c) Anserine
(d) All the above

360. 'Serine deficiency syndrome' is due to a deficiency of:
(a) 3-Phosphoglycerate dehydrogenase
(b) 3-Phosphoserine phosphatase
(c) Either of the above
(d) None of the above

361. The amino acid precursor for selenocysteinyl-tRNA is:
(a) Serine
(b) Methionine
(c) Homocysteine
(d) Glutamate

362. Serine is not a precursor of:
(a) Methionine
(b) Cysteine
(c) Choline
(d) Betaine

363. Thiolysis of α-amino-β-ketobutyrate yields:
(a) Threonine (b) Glycine
(c) Serine (d) Propionyl CoA

364. The following is true for threonine dehydratase catalyzed reaction:
(a) Non-oxidative deamination
(b) Requires vitamin B_6
(c) Forms α-ketobutyrate
(d) All the above

365. Glutamate is a precursor of:
(a) Ornithine (b) Proline
(c) Glutathione (d) All the above

366. The N-methyl-D-aspartate (NMDA) receptor is:
(a) Ligand-gated (b) Voltage-gated
(c) Both (d) None of the above

367. Activation of N-methyl-D-aspartate (NMDA) receptor requires:
(a) Glycine (b) Glutamate
(c) Both (d) None of the above

368. Rapid neuronal death during stroke is due to increased flux of the following ions through N-methyl-D-aspartate (NMDA) receptors:
(a) Na^+ (b) K^+
(c) Ca^{2+} (d) Mg^{2+}

369. In the resting state, N-methyl-D-aspartate (NMDA) receptors are blocked by:
(a) Na^+ (b) K^+
(c) Ca^{2+} (d) Mg^{2+}

370. Synthesis of γ-amino butyric acid (GABA) requires:
(a) Glutamate
(b) Pyridoxal-5-PO_4 (PLP)

(c) Loss of CO_2 from the substrate
(d) All the above

371. 'Umami' (a taste sensation) is attributed to:

(a) Arginine (b) Asparagine
(c) Glutamate (d) Glycine

372. Glutamine synthetase is found in:

(a) Liver and kidneys (b) Brain
(c) Retina (d) All the above

373. In a purine ring, glutamine contributes:

(a) N3 and N9 (b) N1 and N3
(c) N1 and N9 (d) N3 and N7

374. In a pyrimidine ring, aspartate contributes:

(a) C4, C5, C6 and N1 (b) C4, C6 and N1
(c) C5, C6 and N1 (d) C4, C5 and C6

375. The following is true for asparaginase catalyzed reaction:

(a) Yields aspartate
(b) Yields NH_3
(c) The enzyme is administered to patients suffering from asparagine-dependent leukemia
(d) All the above

376. The precursor of tetrahydrobiopterin is:

(a) Biotin (b) GTP
(c) Folic acid (d) Betaine

377. Continued catalysis by phenylalanine hydroxylase requires:

(a) Oxygen (b) Tetrahydrobiopterin
(c) NADPH (d) All the above

378. Humans cannot synthesize:

(a) Vitamin C (b) Aromatic ring
(c) Ergocalciferol (d) All the above

379. Tyrosine aminotransferase is induced by:

(a) Insulin (b) Epinephrine
(c) Cortisol (d) Prolactin

380. Conversion of para-hydroxy phenylpyruvate to homogentisate is an example of:

(a) Oxidation (b) Hydroxylation
(c) Decarboxylation (d) All the above

381. Para-hydroxy phenylpyruvate oxidase requires:

(a) Vitamin C and Cu^{2+} (b) Vitamin C and Fe^{2+}
(c) Vitamin B_6 and Cu^{2+} (d) Vitamin B_6 and Fe^{2+}

382. Homogentisate oxidase requires:

(a) Cu^{2+} (b) Fe^{2+}
(c) Mg^{2+} (d) Zn^{2+}

383. 3,4-Dihydroxyphenyl ring is present in:

(a) Dopamine (b) Norepinephrine
(c) Epinephrine (d) All the above

384. Dopamine β-oxidase does not require:

(a) Oxygen (b) Vitamin C
(c) Fe^{2+} (d) Cu^{2+}

385. Conversion of norepinephrine to epinephrine requires the active form of:

(a) Phenylalanine (b) Tyrosine
(c) Glycine (d) Methionine

386. Vanillyl mandelic acid (VMA) is a catabolite of:

(a) Epinephrine
(b) Insulin
(c) Thyroxine
(d) Vasoactive intestinal polypeptide (VIP)

387. Parkinson's disease is characterized by decreased neuronal synthesis of:

(a) Acetyl choline (b) Dopamine
(c) DOPA (d) Serotonin

388. Tyrosinase requires:

(a) Mg^{2+} (b) Fe^{2+}
(c) Ca^{2+} (d) Cu^{2+}

389. Dihydrobiopterin reductase deficiency leads to:
(a) Type I phenylketonuria (b) Type II phenylketonuria
(c) Albinism (d) Alkaptonuria

390. Phenylketonuria derives its name from increased urinary excretion of:
(a) Phenylalanine (b) Phenyl pyruvate
(c) Phenyl lactate (d) Phenyl acetate

391. In phenylketonuria, the following becomes essential in the diet:
(a) Phenylalanine (b) Alanine
(c) Tyrosine (d) Tryptophan

392. Hepatorenal tyrosinemia is due to a deficiency of:
(a) Tyrosine aminotransferase
(b) Homogentisate oxidase
(c) Tyrosinase
(d) Fumaryl acetoacetate hydrolase

393. Alkaptonuria is due to a deficiency of:
(a) Tyrosine aminotransferase
(b) Homogentisate oxidase
(c) Tyrosinase
(d) Fumaryl acetoacetate hydrolase

394. Albinism is due to a deficiency of:
(a) Tyrosine aminotransferase
(b) Homogentisate oxidase
(c) Tyrosinase
(d) Fumaryl acetoacetate hydrolase

395. Oculocutaneous tyrosinemia is due to a deficiency of:
(a) Tyrosine aminotransferase
(b) Homogentisate oxidase
(c) Tyrosinase
(d) Fumaryl acetoacetate hydrolase

396. Ochronosis is due to a deficiency of:

(a) Tyrosine aminotransferase
(b) Homogentisate oxidase
(c) Tyrosinase
(d) Fumaryl acetoacetate hydrolase

397. Albinism may affect:

(a) Iris and retina
(b) Skin
(c) Hair
(d) All the above

398. Accumulation of carcinogenic alkylating agents may occur due to a deficiency of:

(a) Tyrosine aminotransferase
(b) Homogentisate oxidase
(c) Tyrosinase
(d) Fumaryl acetoacetate hydrolase

399. Intensely dark urine in early postnatal life may be due to a deficiency of:

(a) Tyrosine aminotransferase
(b) Homogentisate oxidase
(c) Tyrosinase
(d) Fumaryl acetoacetate hydrolase

400. Tryptophan pyrrolase is a:

(a) Monooxygenase
(b) Dioxygenase
(c) Dehydrogenase
(d) Decarboxylase

401. Tryptophan pyrrolase is induced by:

(a) Cortisol
(b) Glucagon
(c) Both
(d) None of the above

402. Tryptophan pyrrolase requires:

(a) Fe^{2+}
(b) Cu^{2+}
(c) Zn^{2+}
(d) Mg^{2+}

403. The following is not true for xanthurenic aciduria:

(a) Due to kynureninase deficiency
(b) May be due to vitamin B_6 deficiency

(c) Red urine
(d) Increased formation of kynurenic acid and xanthurenic acid by alternate pathways

404. The following is true for kynurenine hydroxylase:

(a) Requires oxygen
(b) Requires NADPH
(c) Inhibited by estrogen
(d) All the above

405. The following is not seen in vitamin B_6 deficiency:

(a) Increased urinary nicotinamide
(b) Increased urinary N^1-methylnicotinamide
(c) Increased urinary niacin
(d) Decreased synthesis of NMN

406. The following neurotransmitter is not derived from tryptophan:

(a) Serotonin
(b) Kynuramine
(c) Quinolinate
(d) Tyramine

407. The following neurotransmitter is derived from tryptophan:

(a) Melanin
(b) Melatonin
(c) MSH
(d) All the above

408. Increased urinary 5-hydroxy indole acetic acid (5-HIAA) occurs in tumors secreting:

(a) Melanin
(b) Serotonin
(c) Catecholamines
(d) VIP

409. Picolinate is a catabolite of:

(a) Tryptophan
(b) Tyrosine
(c) Threonine
(d) Proline

410. Branched chain amino acid catabolism is most marked in:

(a) Liver
(b) Muscles
(c) Brain
(d) Kidneys

411. The following is true for branched chain amino acid transaminases:

(a) One is Val-specific while the other for Leu/Ile
(b) One is Leu-specific while the other for Val/Ile

(c) One is Ile-specific while the other for Leu/Val
(d) Separate and specific enzyme for each branched chain amino acid

412. The following is true for branched chain keto acid dehydrogenase complex:
(a) Located in inner mitochondrial membrane
(b) Requires TPP
(c) Active in dephosphorylated state
(d) All the above

413. The following is not true for branched chain keto acid dehydrogenase complex:
(a) Catalyzes the reversible oxidative decarboxylation of branched chain keto acids
(b) Requires lipoic acid
(c) Deficiency causes maple syrup urine disease
(d) A multienzyme complex

414. In maple syrup urine disease, the typical odor of urine is due to:
(a) Branched chain amino acids
(b) Branched chain keto acids
(c) Branched chain amino acid acyl CoA
(d) Unsaturated branched chain acyl CoA

415. The following is true for methionine:
(a) Essential amino acid
(b) Excess methionine can be used a source of energy
(c) A lipotropic factor
(d) All the above

416. The following is a methylation reaction with S-adenosyl methionine as a methyl donor except:
(a) Ethanolamine to choline
(b) Guanidoacetate to creatine
(c) N-Acetyl serotonin to melatonin
(d) Epinephrine to norepinephrine

417. In the formation of dimethylglycine from homocysteine, the methyl donor is:

(a) S-adenosyl methionine (b) Betaine
(c) Methyl cobalamine (d) Methyl tetrahydrofolate

418. The following can be derived from homocysteine:

(a) α-Keto butyrate (b) Methionine
(c) Cysteine (d) All the above

419. The following participates in homocysteine metabolism:

(a) Methyl cobalamine (b) Methyl tetrahydrofolate
(c) Pyridoxal-5-PO_4 (PLP) (d) All the above

420. Rhodanese catalyzes:

(a) Detoxification of cyanide to thiocyanate
(b) Formation of thiocysteine
(c) Formation of thiosulphate
(d) Formation of active sulphate

421. Cysteine is a precursor of:

(a) Active sulphate (b) Thiosulphate
(c) Thiocysteine (d) All the above

422. The following can be derived from cysteine except:

(a) Taurine (b) Propionyl CoA
(c) Methionine (d) Bisulphite

423. The following can be derived from cysteine except:

(a) Homocysteine
(b) Disulphide group of proteins
(c) β-Mercaptopyruvate via transamination
(d) Hypotaurine

424. Active sulphate is:

(a) 3′-Phosphoadenosine-3′-phosphosulphate
(b) 3′-Phosphoadenosine-5′-phosphosulphate
(c) 5′-Phosphoadenosine-5′-phosphosulphate
(d) 5′-Phosphoadenosine-3′-phosphosulphate

425. Type I homocysteinuria is due to a deficiency of:

(a) Cystathionine β-synthase
(b) N^5,N^{10}-Methylene tetrahydrofolate reductase
(c) N^5-Methyl tetrahydrofolate:homocysteine transmethylase
(d) Intestinal absorption of cobalamine

426. Type II homocysteinuria is due to a deficiency of:

(a) Cystathionine β-synthase
(b) N^5,N^{10}-Methylene tetrahydrofolate reductase
(c) N^5-Methyl tetrahydrofolate:homocysteine transmethylase
(d) Intestinal absorption of cobalamine

427. Type III homocysteinuria is due to a deficiency of:

(a) Cystathionine β-synthase
(b) N^5,N^{10}-Methylene tetrahydrofolate reductase
(c) N^5-Methyl tetrahydrofolate:homocysteine transmethylase
(d) Intestinal absorption of cobalamine

428. Type IV homocysteinuria is due to a deficiency of:

(a) Cystathionine β-synthase
(b) N^5,N^{10}-Methylene tetrahydrofolate reductase
(c) N^5-Methyl tetrahydrofolate : homocysteine transmethylase
(d) Intestinal absorption of cobalamine

429. Hyperhomocysteinemia is:

(a) Congenital
(b) Acquired
(c) Either of the above
(d) Does not exists

430. The following may be associated with acquired hyperhomocysteinemia:

(a) Atherosclerosis
(b) Alzheimer's disease
(c) Osteoporosis
(d) Any of the above

431. The following may be associated with acquired hyperhomocysteinemia except:

(a) Increased formation of homocysteine thiolactone
(b) Increased formation of oxidized-LDL

(c) Increased serum folate, vitamin B_{12} and vitamin B_6 in old age
(d) Increased generation of free radicals

432. Cystinosis may involve:

(a) Kidneys
(b) Bone marrow
(c) Cornea
(d) All the above

433. The two amino groups of lysine are digested as:

(a) α and β
(b) α and γ
(c) α and δ
(d) α and ε

434. The following is true for α-amino adipic semialdehyde synthase:

(a) A bifunctional enzyme
(b) Deficiency results in hyperlysinuria
(c) First enzyme in lysine catabolism
(d) All the above

435. The following is not true for lathyrism:

(a) Linked to increased consumption of 'khesari dal' (*Lathyrus sativus*)
(b) Causative toxin is β-oxalyl amino alanine (BOAA)
(c) Inappropriate collagen maturation
(d) Treatment requires restriction of vitamin C

436. In familial lysinuric protein intolerance, plasma level of the following is not reduced:

(a) Ammonia
(b) Lysine
(c) Ornithine
(d) Arginine

437. In familial lysinuric protein intolerance, the following membrane transport system is defective:

(a) SLC7A7
(b) ATP7A
(c) Group translocation
(d) Ornithine-citrulline antiport

438. In the small intestine, citrulline is synthesized from:
(a) Arginine (b) Glutamate
(c) Glutamine (d) Ornithine

439. In the kidneys, arginine is synthesized from:
(a) Ornithine (b) Glutamate
(c) Citrulline (d) Glutamine

440. The following is derived from arginine:
(a) Proline (b) Glutamate
(c) Ornithine (d) All the above

441. Nitric oxide synthase requires the following except:
(a) NADH (b) Tetrahydrobiopterin
(c) Heme (d) NADPH

442. Nitric oxide synthase requires the following except:
(a) FAD (b) FMN
(c) NADH (d) NADPH

443. The precursor of agmatine (a physiologically active antihypertensive) is:
(a) Arginine (b) Histidine
(c) Glutamate (d) Aspartate

444. In nitric oxide synthase, the following three prosthetic groups are located in a single polypeptide chain:
(a) Tetrahydrobiopterin, heme and FAD
(b) Tetrahydrobiopterin, heme and FMN
(c) FMN, FAD and Heme
(d) Tetrahydrobiopterin, FMN and FAD

445. The following is not Ca^{2+}-dependent:
(a) nNOS (b) iNOS
(c) eNOS (d) None of the above

446. The vasoconstrictor function of endothelin is mediated through:
(a) Endothelin A receptors (b) Endothelin B receptors
(c) Both (d) None of the above

447. The second messenger for nitric oxide is:

(a) cAMP
(b) cGMP
(c) Ca^{2+}
(d) Diacylglycerol (DAG)

448. The following is not true for polyamine synthesis from ornithine:

(a) Requires S-adenosyl methionine
(b) Reaction proceeds through ornithine to putrescine to spermidine to spermine
(c) 5′-Methylthioadenosine is a side-product
(d) S-Adenosyl methionine donates its methyl group during the reactions

449. The following is a polyamine:

(a) Putrescine
(b) Spermidine
(c) Spermine
(d) All the above

450. The following is not true for polyamines:

(a) Ornithine is the precursor
(b) Highly anionic
(c) Bind to DNA
(d) Can be synthesized by intestinal flora

451. The following is not true for histamine:

(a) Synthesized from histidine by decarboxylation
(b) A vasoconstrictor
(c) Synthesis requires Pyridoxal-5-PO_4 (PLP)
(d) Responsible for allergic manifestations

452. The following is not true for histidinemia:

(a) Due to histidase deficiency
(b) Speech defects
(c) Slowed mental development
(d) Increased urocanate in sweat

453. Large amounts of the following amino acid may be normally detected in the urine during pregnancy:

(a) Proline
(b) Alanine
(c) Methionine
(d) Histidine

454. Lohmann's reaction is catalyzed by:
(a) Arginine glycine transamidinase
(b) Methyl transferase
(c) Creatine phosphokinase
(d) A spontaneous (non-enzymatic) reaction

455. The conversion of creatine phosphate to creatinine is catalyzed by:
(a) Arginine glycine transamidinase
(b) Methyl transferase
(c) Creatine phosphokinase
(d) A spontaneous (non-enzymatic) reaction

456. The conversion of creatine phosphate to creatinine involves:
(a) Dehydration
(b) Dephosphorylation
(c) Cyclization
(d) All the above

457. Daily urinary creatinine excretion is:
(a) 0.1–0.5 g
(b) 0.5–1.0 g
(c) 1.0–1.5 g
(d) 1.5–2.0 g

458. The synthesis of glutathione is regulated mainly by the availability of:
(a) Glycine
(b) Cysteine
(c) Glutamate
(d) All the above

459. Glutathione participates in:
(a) Rearrangement of protein disulphide bonds
(b) Synthesis of leukotrienes
(c) Group translocation of amino acids
(d) All the above

460. The following is not a known function of glutathione:
(a) Fibrin stabilization
(b) Lipid-soluble antioxidant
(c) Conjugation of some drugs
(d) Part of 'glucose tolerance factor'

461. The following is not a fate of glucose-6-phosphate in the fed state:

(a) Glycogenesis (b) Glycolysis
(c) HMP shunt (d) Gluconeogenesis

462. In a healthy 70 kg man, the fat content in adipose tissue is:

(a) 5 kg (b) 15 kg
(c) 25 kg (d) 35 kg

463. The following is not true for anorexia nervosa:

(a) Mainly in adolescent girls
(b) May be associated with early menarche
(c) Patients never develop bulimia nervosa
(d) May be life-threatening

464. During starvation, blood levels of the following increases:

(a) Carnitine (b) Free fatty acids
(c) Ketone bodies (d) All the above

465. The preferred fuel substrate for adipose tissue is:

(a) Glucose (b) Fatty acids
(c) Amino acids (d) Ketone bodies

466. The preferred fuel substrate for brain is:

(a) Glucose (b) Fatty acids
(c) Amino acids (d) Ketone bodies

467. The preferred fuel substrate for resting skeletal muscle is:

(a) Glucose (b) Fatty acids
(c) Amino acids (d) Ketone bodies

468. The preferred fuel substrate for exercising skeletal muscle is:

(a) Glucose (b) Fatty acids
(c) Amino acids (d) Ketone bodies

469. During exercise, the following is released from muscles into the circulation:

(a) Glucose (b) Ketone bodies
(c) Lactate (d) Glycerol

470. During starvation, the following is released from muscles into the circulation:

(a) Glucose
(b) Ketone bodies
(c) Glycerol
(d) Branched chain amino acids

471. Fischer ratio is:

(a) Branched chain amino acids: Aromatic amino acids
(b) Branched chain amino acids: Sulphur containing amino acids
(c) Aromatic amino acids: Sulphur containing amino acids
(d) Acidic amino acids: Basic amino acids

472. An altered plasma branched chain amino acids : aromatic amino acids ratio may be associated with:

(a) Hepatic cirrhosis
(b) Tardive dyskinesia
(c) Insulin resistance
(d) All the above

473. At physiological pH, nucleic acids are:

(a) Negatively charged
(b) Positively charged
(c) Uncharged
(d) Uncharged zwitterions

474. At physiological pH, guanine exists as:

(a) Lactim (enol)
(b) Lactam (keto)
(c) Lactim (keto)
(d) Lactam (enol)

475. At physiological pH, thymine exists as:

(a) Lactim (enol)
(b) Lactam (keto)
(c) Lactim (keto)
(d) Lactam (enol)

476. The following are vasodilators except:

(a) Prostacyclin
(b) Adenosine
(c) Nitric oxide
(d) Endothelin

477. The following is not true for adenosine:

(a) A local hormone
(b) Promotes sleep
(c) A vasodilator
(d) Used to treat supraventricular tachycardia

478. In the Watson-Crick DNA model, the helix pitch is:

(a) 0.34 nm (b) 34 nm
(c) 3.4 nm (d) 340 nm

479. In the Watson-Crick DNA model, the number of base-pairs per turn are:

(a) 3 (b) 10
(c) 13 (d) 20

480. In the Watson-Crick DNA model, the N-base and sugar-phosphate moieties respectively occupy the following positions:

(a) Core and periphery (b) Periphery and core
(c) Core only (d) Periphery only

481. According to Chargaff's rule, the following is true in a DNA double helix:

(a) A = T (b) G = C
(c) Both (d) None of the above

482. In the Watson-Crick DNA model, the number of H-bond(s) between A and T is/are:

(a) 1 (b) 2
(c) 3 (d) 4

483. In the Watson-Crick DNA model, the number of H-bond(s) between G and C is/are:

(a) 1 (b) 2
(c) 3 (d) 4

484. During DNA denaturation, the following bond(s) is/are broken:

(a) Phosphodiester (b) Hydrogen
(c) Glycosidic (d) All the above

485. The following occurs during DNA denaturation:

(a) Decrease in viscosity
(b) Strand separation
(c) Increase in UV absorbance
(d) All the above

486. The melting temperature of DNA depends on the following:
(a) Nature of solvent
(b) Ionic strength and pH of the solution
(c) Base composition of DNA
(d) All the above

487. The Watson-Crick model originally described the following:
(a) Z-DNA
(b) B-DNA
(c) A-DNA
(d) Denatured DNA

488. The helix pitch is least in:
(a) B-DNA
(b) A-DNA
(c) Z-DNA
(d) Equal in all

489. The major groove is narrowest in:
(a) B-DNA
(b) A-DNA
(c) Z-DNA
(d) Equal in all

490. The minor groove is narrowest in:
(a) B-DNA
(b) A-DNA
(c) Z-DNA
(d) Equal in all

491. The helix rotation is left-handed in:
(a) B-DNA
(b) A-DNA
(c) Z-DNA
(d) All the above

492. The number of base-pairs per turn are least in:
(a) B-DNA
(b) A-DNA
(c) Z-DNA
(d) Equal in all

493. The biologically commonest form of DNA is:
(a) B-DNA
(b) A-DNA
(c) Z-DNA
(d) None of the above

494. The diameter of Watson-Crick B-DNA double helix is:
(a) 0.2 nm
(b) 2.0 nm
(c) 0.34 nm
(d) 3.4 nm

495. DNA serves as a template for:

(a) Transcription (b) Replication
(c) Both (d) None of the above

496. The following is not true for plasmids:

(a) Small
(b) Circular
(c) Extrachromosomal
(d) Have lost the power to replicate

497. The following is not true for RNA:

(a) Acid labile
(b) Contain uracil
(c) Contain ribose
(d) For single stranded RNA, Chargaff's rule is not applicable

498. Chargaff's rule is applicable to:

(a) mRNA (b) tRNA and rRNA
(c) dsDNA (d) All the above

499. hnRNA is found in the:

(a) Nucleus
(b) Cytoplasm
(c) Ribosomes
(d) Smooth endoplasmic reticulum

500. The following is not true for tRNA:

(a) Clover leaf-like structure
(b) Contains several unusual bases
(c) Also called insoluble RNA
(d) Extensive intra-molecular base-pairing

501. The specificity of a tRNA molecule lies in the:

(a) Amino acid acceptor arm (b) Anticodon arm
(c) TψC arm (d) DHU arm

502. Unusual bases are present in the following structural part of tRNA:

(a) Amino acid acceptor arm (b) Anticodon arm
(c) DHU arm (d) Extra arm

503. The modified base in DHU arm of tRNA is:

(a) Dihydrouracil (b) Pseudouracil
(c) 5-Fluorouracil (d) None of the above

504. 80% of total cellular RNA is present in the:

(a) Cytosol (b) Ribosomes
(c) Nucleoplasm (d) Nucleolus

505. Ribosomes contain:

(a) rRNA (b) mRNA
(c) tRNA (d) All the above

506. Intact mammalian ribosomes have a sedimentation velocity of:

(a) 80S (b) 60S
(c) 40S (d) 100S

507. C6 of a purine ring is contributed by:

(a) Glycine
(b) Aspartate
(c) Bicarbonate
(d) N^{10}-Formyl tetrahydrofolate

508. The following is true for PRPP:

(a) Precursor for purines
(b) Precursor for pyrimidines
(c) Used in the purine salvage pathway
(d) All the above

509. The first purine nucleotide synthesized *de novo* is:

(a) AMP (b) GMP
(c) IMP (d) None of the above

510. C2 and C8 of a purine ring are contributed by:

(a) N^{10}-Formyl tetrahydrofolate
(b) N^5,N^{10}-Methylene tetrahydrofolate
(c) N^5-Methyl tetrahydrofolate
(d) N^5-Formyl tetrahydrofolate

511. Number of moles of ATP used to synthesize one mole of IMP are:

(a) 2
(b) 4
(c) 6
(d) 8

512. Purine nucleotides are synthesized in:

(a) Cytosol
(b) Mitochondria
(c) Nucleus
(d) Partly in all the above

513. Glutamine PRPP amidotransferase is negatively allosterically regulated by the following except:

(a) PRPP
(b) IMP
(c) GMP
(d) AMP

514. The following catalyzes the committed step in purine nucleotide biosynthesis:

(a) PRPP synthetase
(b) Phosphoribosylamine synthetase
(c) Transformylase
(d) Phosphoribosyl glycinamide synthetase

515. The following amphibolic intermediate is obtained during *de novo* purine synthesis:

(a) Succinate
(b) Citrate
(c) Oxaloacetate
(d) Fumarate

516. The main source of adenine in the purine salvage pathway is:

(a) Polyamines
(b) Degradation of purine nucleotides
(c) *De novo* adenine synthesis
(d) None of the above

517. The following is true for adenine phosphoribosyl transferase (APRTase):

(a) Involved in purine salvage pathway
(b) Requires PRPP
(c) Competitively inhibited by AMP
(d) All the above

518. The following is true for hypoxanthine guanine phosphoribosyl transferase (HGPRTase):

(a) Involved in purine salvage pathway
(b) Requires PRPP
(c) Competitively inhibited by IMP and GMP
(d) All the above

519. The following is true for HGPRTase except:

(a) Gene is located on X chromosome
(b) Deficiency disease affects females only
(c) Salvages hypoxanthine and guanine
(d) Regulated by end products

520. The following may be seen in Lesch Nyhan syndrome:

(a) Hyperuricemia (b) Mental retardation
(c) Self mutilation (d) All the above

521. The following may be seen in Lesch Nyhan syndrome except:

(a) Increased intracellular PRPP
(b) Decreased intracellular IMP
(c) Decreased intracellular GMP
(d) Decreased *de novo* purine synthesis

522. The following are salvaged in the purine salvage pathway:

(a) Free purine bases (b) Purine nucleosides
(c) Both (d) None of the above

523. A nucleoside kinase reaction participates in:

(a) *De novo* purine nucleotide synthesis
(b) Purine salvage
(c) Purine catabolism
(d) All the above

524. The main advantage of purine salvage pathway is:

(a) Not regulated by end products
(b) Does not require PRPP
(c) Conserves cellular energy
(d) All the above

525. Phosphoribosylamine synthetase (glutamine PRPP amidotransferase) is absent in:

(a) Hepatocytes (b) Enterocytes
(c) Erythrocytes (d) Neutrophils

526. The following is not true for AMP deaminase reaction:

(a) Forms IMP (b) Liberates NH_4^+
(c) Inhibited by Pi (d) Inhibited by K^+

527. The following is not true for AMP deaminase reaction:

(a) Activated by ATP (b) Inhibited by Pi
(c) Inhibited by GDP (d) Activated by GTP

528. Adenosine deaminase (ADA) activity is highest in:

(a) Muscles (b) Liver
(c) Kidneys (d) Brain

529. The substrate for ADA is:

(a) Adenosine (b) Deoxyadenosine
(c) Both (d) None of the above

530. The substrate for purine nucleoside phosphorylase (PNP) is:

(a) Inosine (b) Guanosine
(c) Both (d) None of the above

531. The substrate for xanthine oxidase is:

(a) Hypoxanthine (b) Xanthine
(c) Both (d) None of the above

532. The following is not true for the xanthine oxidase catalyzed reaction:

(a) Requires oxygen
(b) Generates H_2O_2
(c) Enzyme is a dimeric protein
(d) Absent in liver and small intestine

533. Xanthine oxidase contains:

(a) FAD (b) Molybdenum (Mo)
(c) FeS (d) All the above

534. The following is not true for Severe Combined Immunodeficiency (SCID):

(a) Deficiency of adenosine deaminase
(b) Decreased cellular dATP
(c) Increased cellular 2′-deoxyadenosine
(d) Decreased cellular ATP

535. The following is not a known treatment modality for SCID:

(a) Bone marrow transplantation
(b) Plasmapheresis
(c) ADA-Polyethylene glycol replacement
(d) ADA gene therapy

536. SCID is characterized by:

(a) Decreased T cell function but normal B cell function
(b) Decreased T and B cell functions
(c) Normal T cell function but decreased B cell function
(d) Normal T and B cell functions

537. PNP deficiency is characterized by:

(a) Decreased T cell function but normal B cell function
(b) Decreased T and B cell functions
(c) Normal T cell function but decreased B cell function
(d) Normal T and B cell functions

538. Decreased uric acid synthesis is seen in:

(a) SCID
(b) PNP deficiency
(c) Both
(d) None of the above

539. Increased uric acid synthesis occurs in the following except:

(a) ADA deficiency
(b) HGPRTase deficiency
(c) Increased activity of PRPP synthetase
(d) Glucose-6-phosphatase deficiency

540. In glucose-6-phosphatase deficiency, hyperuricemia is due to:

(a) Increased uric acid synthesis
(b) Decreased uric acid excretion

(c) Both
(d) None of the above

541. In chronic alcoholism, hyperuricemia is due to:
(a) Increased uric acid synthesis
(b) Decreased uric acid excretion
(c) Both
(d) None of the above

542. In diabetic ketoacidosis, hyperuricemia is due to:
(a) Increased uric acid synthesis
(b) Decreased uric acid excretion
(c) Both
(d) None of the above

543. In tuberculosis patients on pyrazinamide, hyperuricemia is due to:
(a) Increased uric acid synthesis
(b) Decreased uric acid excretion
(c) Both
(d) None of the above

544. Allopurinol decreases:
(a) *De novo* purine nucleotide synthesis
(b) Uric acid synthesis
(c) Both
(d) None of the above

545. In a pyrimidine ring, bicarbonate contributes:
(a) C2
(b) C4
(c) C5
(d) C6

546. The following enzyme catalyzes the committed step in pyrimidine nucleotide biosynthesis:
(a) PRPP synthetase
(b) Carbamoylphosphate synthase-II (CPS-II)
(c) Aspartate transcarbamoylase
(d) Carbamoylphosphate synthase-I (CPS-I)

547. The following is not a part of a single trifunctional protein:

(a) Carbamoylphosphate synthase-II (CPS-II)
(b) Aspartate transcarbamoylase
(c) Dihydroorotase
(d) Dihydroorotate dehydrogenase

548. The following is a bifunctional enzyme:

(a) Carbamoylphosphate synthase-I (CPS-I)
(b) Carbamoylphosphate synthase-II (CPS-II)
(c) Uridine monophosphate (UMP) synthase
(d) Dihydroorotate dehydrogenase

549. The following is not true for Uridine monophosphate (UMP) synthase:

(a) Has orotate phosphoribosyl transferase and orotate-monophosphate (OMP) decarboxylase activities
(b) Deficiency results in orotic aciduria
(c) Deficiency results in severe anemia
(d) Treatment of deficiency disease is based on dietary uridine restriction

550. The following is a mitochondrial enzyme in pyrimidine biosynthesis:

(a) Carbamoylphosphate synthase-II (CPS-II)
(b) Aspartate transcarbamoylase
(c) Dihydroorotase
(d) Dihydroorotate dehydrogenase

551. The following is synthesized first during pyrimidine nucleotide biosynthesis:

(a) UMP (b) CMP
(c) TMP (d) None of the above

552. The amino group donor in the CTP synthetase reaction is:

(a) Asparagine (b) Aspartate
(c) Glutamine (d) Glutamate

553. The methyl group donor in thymidylate synthesis is:

(a) S-Adenosyl methionine
(b) Betaine
(c) N^5, N^{10}-Methylene tetrahydrofolate
(c) N^5-Methyl tetrahydrofolate

554. 5-Fluoro deoxyuridylate (FdUMP) inhibits:

(a) Thymidylate synthase
(b) Dihydrofolate reductase
(c) Both
(d) None of the above

555. Methotrexate inhibits:

(a) Thymidylate synthase
(b) Dihydrofolate reductase
(c) Both
(d) None of the above

556. The following is a suicide substrate:

(a) 5-Fluoro deoxyuridylate (FdUMP)
(b) Allopurinol
(c) Both
(d) None of the above

557. The following is an 'antifolate':

(a) Amethopterin
(b) Aminopterin
(c) Trimethoprim
(d) All the above

558. Dihydrofolate reductase is inhibited by the following except:

(a) Amethopterin
(b) Aminopterin
(c) Trimethoprim
(d) Folinic acid

559. CTP inhibits:

(a) CPS-II
(b) CTP synthetase
(c) Both
(d) None of the above

560. The following nitrogenous base is not salvaged:

(a) Cytosine
(b) Orotate
(c) Uracil
(d) Thymine

561. The following is measured in urine to assess DNA turnover:

(a) β-Alanine
(b) Putrescine
(c) β-Amino isobutyric acid
(d) Adenosine deaminase

562. Ribonucleotide reductase requires the following:

(a) Thioredoxin (b) Glutaredoxin
(c) Either of the above (d) None of the above

563. Mammalian ribonucleotide reductase is a:

(a) Monomer (b) Homodimer
(c) Heterodimer (d) Tetramer

564. The prosthetic group in thioredoxin reductase is:

(a) Heme (b) FMN
(c) FAD (d) GSH

565. The phosphoryl donor in nucleoside diphosphate kinase reaction may be:

(a) dATP (b) ATP
(c) Either of the above (d) None of the above

566. 'Folinic acid rescue' is employed during:

(a) Sickle cell crisis (b) Methotrexate chemotherapy
(c) Warfarin overdose (d) Scurvy

567. The following is true for sodium urate:

(a) Poor aqueous solubility
(b) Precipitates in acidic pH
(c) Dehydration promotes precipitation
(d) All the above

568. Uric acid may deposit in:

(a) Synovial fluid (b) Renal interstitium
(c) Ureter (d) All the above

569. Needle-shaped crystals with strong negative birefringence under polarized microscope are diagnostic of:

(a) Cystine (b) Monosodium urate
(c) Triple phosphate (d) Cholesterol

570. 'Podagra' generally refers to:

(a) Pellagra with all the classical signs/symptoms
(b) Classical phenylketonuria

(c) Classical uric acid arthropathy of great toe
(d) Passive immunization

571. Catabolism of GTP produces:
(a) Neopterin (b) Dihydroneopterin
(c) Tetrahydrobiopterin (d) All the above

572. Increased plasma neopterin levels may suggest:
(a) Ongoing oxidative stress
(b) Exposure to silica
(c) Early phase of cellular immune response
(d) All the above

573. The chemical workshop of the body is:
(a) Brain (b) Heart
(c) Liver (d) Kidney

574. The following proteins are not synthesized by hepatocytes:
(a) α_1-Globulins (b) α_2-Globulins
(c) γ-Globulins (d) β-Globulins

575. The following metabolic pathway does not occur in the liver:
(a) Glycolysis
(b) Tryptophan catabolism
(c) Purine nucleotide synthesis
(d) Ketolysis

576. The liver participates in the synthesis of:
(a) Insulin-like growth factor (IGF)
(b) T_3
(c) Vitamin D
(d) All the above

577. The following is degraded in the liver:
(a) Insulin (b) Glucagon
(c) Growth hormone (d) All the above

578. The kidneys do not synthesize:

(a) Glucose (b) Glutathione
(c) Urea (d) Ammonia

579. 'Overflow proteinuria' occurs in:

(a) Nephrotic syndrome (b) Multiple myeloma
(c) Chronic pyelonephritis (d) All the above

580. Cystatin C is produced by:

(a) Cells lining the proximal convoluted tubules only
(b) All nucleated cells
(c) RBC only
(d) Juxtaglomerular cells only

581. The following is secreted in the nephrons:

(a) H^+ (b) Urea
(c) Creatinine (d) All the above

582. The mean standard urea clearance is:

(a) 54 ml/min (b) 75 ml/min
(c) 100 ml/min (d) 125 ml/min

583. The maximum urea clearance is:

(a) 54 ml/min (b) 75 ml/min
(c) 100 ml/min (d) 125 ml/min

584. The normal inulin clearance is:

(a) 54 ml/min (b) 75 ml/min
(c) 100 ml/min (d) 125 ml/min

585. In a neonate, creatinine clearance of 35 ml/min most likely indicates:

(a) Bilateral stricture of ureters
(b) Stricture urethra
(c) Polycystic kidney disease
(d) Normal finding

586. The filtration fraction is calculated by:

(a) Inulin clearance ÷ PAH clearance
(b) PAH clearance ÷ Inulin clearance
(c) Inulin clearance – PAH clearance
(d) PAH clearance – Inulin clearance

587. A useful index of renal plasma flow is:

(a) Inulin clearance (b) PAH clearance
(c) Creatinine clearance (d) Urea clearance

588. Serum creatinine may be increased by dietary intake of:

(a) Pulses (b) Soybean
(c) Milk (d) Meat

589. 'Sweaty feet' odour of urine suggests the presence of:

(a) Isovaleric acid (b) Glutaric acid
(c) Butyric acid (d) Any of the above

590. For a positive urinary nitrite test, urine must be retained in the urinary bladder for approximately:

(a) 4 min (b) 40 min
(c) 1 h (d) 4 h

591. Under polarized light, the following urinary constituent gives a 'Maltese cross' appearance:

(a) Cholesterol (b) Phosphate
(c) Urate (d) Oxalate

592. The following urinary crystals microscopically appear either as rhomboid prisms or amorphous:

(a) Uric acid (b) Calcium oxalate
(c) Cystine (d) Phosphate

593. The following urinary crystals microscopically appear either as bipyramidal or envelope-shaped:

(a) Uric acid (b) Calcium oxalate
(c) Cystine (d) Phosphate

594. The following urinary crystals microscopically appear as hexagonal:

(a) Uric acid
(b) Calcium oxalate
(c) Cystine
(d) Phosphate

595. The following urinary crystals microscopically appear as coffin-lids (bevelled rectangular prisms):

(a) Uric acid
(b) Calcium oxalate
(c) Cystine
(d) Triple phosphate

596. The following urinary crystals microscopically appear as thorn apples:

(a) Ammonium biurate
(b) Calcium oxalate
(c) Cystine
(d) Triple phosphate

597. The following generally crystallizes in acidic urine:

(a) Uric acid
(b) Calcium oxalate
(c) Both
(d) None of the above

598. The following is increasingly excreted in urine in Fanconi's syndrome:

(a) Glucose
(b) Phosphate
(c) Amino acids
(d) All the above

599. The following is not true for ornithine transaminase deficiency:

(a) Increased plasma ornithine
(b) Gyrate atrophy of retina with tunnel vision
(c) May result in blindness
(d) Treated by increasing dietary arginine

600. The following is true for apolipoprotein J:

(a) Also called clusterin
(b) Increased blood levels are associated with dementia
(c) May be used to predict the future onset of Alzheimer's disease
(d) All the above

SECTION-3

SPECIALIZED PROTEINS: ENZYMES, ELECTRON TRANSPORT, OXIDATIVE PHOSPHORYLATION, HEMO-PROTEINS, MUSCLE CONTRACTION

1. The following are tetramers except:
(a) Phosphorylase
(b) Hexokinase
(c) Fructose 1,6-bisphosphatase
(d) Lactate dehydrogenase

2. The following are dimers except:
(a) Phosphofructokinase
(b) Creatine phosphokinase
(c) Fructose 1,6-bisphosphatase
(d) Hexokinase

3. The following are multienzyme complexes with 3 different enzyme activities except:
(a) Fatty acid synthase
(b) Pyruvate dehydrogenase
(c) α-Ketoglutarate dehydrogenase
(d) Branched chain keto acid dehydrogenase

4. The enzyme nomenclature 1.1.1.1. stands for:
(a) Aldehyde dehydrogenase (b) Alcohol dehydrogenase
(c) Isocitrate dehydrogenase (d) Pyruvate dehydrogenase

5. The following is not a lyase:

(a) Glucose-6-phosphatase (b) Aldolase
(c) Argininosuccinase (d) Histidase

6. The cofactor for pyruvate kinase is:

(a) Cl^- (b) K^+
(c) Ca^{2+} (d) Cu^{2+}

7. The following is not true for coenzymes:

(a) Thermolabile (b) Dialyzable
(c) Low M_r (d) Organic

8. The following is not true for prosthetic group:

(a) Thermostable (b) Dialyzable
(c) Low M_r (d) Loosely bound

9. Hexokinase phosphorylates:

(a) Glucose (b) Fructose
(c) Mannose (d) All the above

10. Glucokinase phosphorylates:

(a) Glucose (b) Fructose
(c) Mannose (d) All the above

11. Hexokinase does not phosphorylate:

(a) Glucose (b) Fructose
(c) Mannose (d) Galactose

12. A substrate binds to an enzyme through the following except:

(a) Irreversible covalent linkage
(b) H-bonds
(c) Electrostatic bonds
(d) Hydrophobic interactions

13. Transamination is an example of a multisubstrate reaction showing:

(a) Ping-pong mechanism (b) Sequential mechanism
(c) Both (d) None of the above

14. The following is a rigid model of enzyme action:

(a) Substrate strain (b) Lock and key
(c) Induced-fit (d) None of the above

15. In enzyme catalysis, free energy of activation refers to the energy difference between:

(a) Transition state and product
(b) Substrate and product
(c) Substrate and transition state
(d) A pair of substrates

16. In enzyme catalysis, the following is reduced:

(a) Free energy of substrate
(b) Free energy of transition state
(c) Free energy of product
(d) All the above

17. Aced-base catalysis by RNaseA is attributed to:

(a) His12 (b) His119
(c) Both (d) None of the above

18. During acid-base catalysis, the following may act as an acid:

(a) –SH group of cysteine (b) –OH group of tyrosine
(c) ε-Amino group of lysine (d) Any of the above

19. Covalent catalysis is also called:

(a) Acid-base catalysis (b) Nucleophilic catalysis
(c) Metal ion catalysis (d) Any of the above

20. Optimum pH for pepsin is:

(a) 1.0 (b) 2.0
(c) 3.0 (d) 4.0

21. When the maximum velocity of an enzyme catalyzed reaction is independent of substrate concentration, the reaction order is:

(a) First (b) Second
(c) Zero (d) None of the above

22. K_m is:

(a) Directly proportional to enzyme concentration
(b) Inversely proportional to enzyme concentration
(c) Half of enzyme concentration
(d) Independent of enzyme concentration

23. The slope of a Lineweaver-Burk plot is given by:

(a) $1/V_{max}$
(b) $-K_m$
(c) $-K_m/V_{max}$
(d) $-V_{max}/K_m$

24. The following is not true for hexokinase:

(a) Low K_m
(b) High affinity for glucose
(c) More important during fed state
(d) Not specific for glucose

25. The following is not true for glucokinase:

(a) High K_m
(b) Low affinity for glucose
(c) More important during fasting state
(d) Specific for glucose

26. The following is not true for K_m:

(a) Constant for a given enzyme
(b) A measure of enzyme-substrate affinity
(c) Can be experimentally determined
(d) For a given enzyme, the value never differs from person to person

27. The following is true for high K_m aldehyde dehydrogenase:

(a) Oxidizes acetaldehyde slowly
(b) Person more prone to flushing and headache following alcohol intake
(c) Both
(d) None of the above

28. Sulphonamides inhibit dihydropteroate synthase by:

(a) Competitive inhibition
(b) Non-competitive inhibition

(c) Uncompetitive inhibition
(d) Allosteric inhibition

29. Dicumarol inhibits epoxide reductase by:
(a) Competitive inhibition
(b) Non-competitive inhibition
(c) Uncompetitive inhibition
(d) Allosteric inhibition

30. 5-Fluorouracil inhibits thymidylate synthetase by:
(a) Competitive inhibition (b) Non-competitive inhibition
(c) Uncompetitive inhibition (d) Allosteric inhibition

31. Phenylalanine inhibits alkaline phosphatase by:
(a) Competitive inhibition (b) Non-competitive inhibition
(c) Uncompetitive inhibition (d) Allosteric inhibition

32. Competitive inhibition is characterized by:
(a) Unchanged V_{max} and increased (apparent) K_m
(b) Decreased V_{max} and unchanged K_m
(c) Decreased V_{max} and decreased (apparent) K_m
(d) None of the above

33. Non-competitive inhibition is characterized by:
(a) Unchanged V_{max} and increased (apparent) K_m
(b) Decreased V_{max} and unchanged K_m
(c) Decreased V_{max} and decreased (apparent) K_m
(d) None of the above

34. Uncompetitive inhibition is characterized by:
(a) Unchanged V_{max} and increased (apparent) K_m
(b) Decreased V_{max} and unchanged K_m
(c) Decreased V_{max} and decreased (apparent) K_m
(d) None of the above

35. Irreversible inhibition is characterized by:
(a) Unchanged V_{max} and increased (apparent) K_m
(b) Decreased V_{max} and unchanged K_m
(c) Decreased V_{max} and decreased (apparent) K_m
(d) None of the above

36. Allopurinol is a:
(a) Competitive inhibitor of xanthine oxidase
(b) A substrate for xanthine oxidase
(c) Both
(d) None of the above

37. Mechanism-based enzyme inhibition is also called:
(a) Suicide inhibition
(b) Trojan horse substrate inhibition
(c) Both
(d) None of the above

38. The following is not an inducible enzyme:
(a) Hexokinase
(b) HMG CoA reductase
(c) Tyrosine aminotransferase
(d) Tryptophan pyrrolase

39. ADP is not an allosteric activator of:
(a) Hexokinase
(b) Isocitrate dehydrogenase
(c) Glutamate dehydrogenase
(d) Pyruvate carboxylase

40. Covalent modification of enzyme activity commonly occurs through:
(a) Serine
(b) Threonine
(c) Tyrosine
(d) All the above

41. The following is a monomeric allosteric enzyme:
(a) Ribonucleoside diphosphate reductase
(b) Pyruvate UDP-N-Acetyl glucosamine transferase
(c) Both
(d) None of the above

42. The following is not true for isoenzymes:
(a) Differ in substrate specificity
(b) Differ in structure

(c) Differ in electrophoretic mobility
(d) Differ in immunological properties

43. Isoenzymes may be present in:

(a) Different tissues
(b) Different cell types of a tissue
(c) Different subcellular compartments of a cell
(d) All the above

44. The following is not true for alkaline phosphatase (ALP):

(a) Its isoenzymes have identical carbohydrate content
(b) Monomeric
(c) α_1-ALP is present in liver
(d) Increased serum α_1-ALP is seen in obstructive jaundice

45. The following is not true for placental ALP:

(a) An α_2-ALP
(b) Heat labile
(c) Contains sialic acid
(d) Increased serum level is seen in pregnancy

46. A useful serum marker is some cases of bronchogenic carcinoma is:

(a) γ-Alkaline phosphatase (γ-ALP)
(b) Regan ALP
(c) LDH-2
(d) LDH-5

47. LDH-5 is present mainly in:

(a) Myocardium (b) RBC
(c) Liver and muscles (d) Brain

48. In acute myocardial infarction, a flipped ratio refers to:

(a) LDH-2>LDH-1 (b) LDH-1>LDH-2
(c) LDH-1>CPK-MB (d) LDH-2>CPK-MB

49. In acute myocardial infarction, the marker to rise earliest in serum is:

(a) CPK-MM (b) LDH-1
(c) CPK-MB (d) ALT (SGPT)

50. The following is a normal finding:

(a) Increased serum alkaline phosphatase (ALP) in third trimester of pregnancy
(b) Increased serum alkaline phosphatase (ALP) in growing children
(c) Both
(d) None of the above

51. One enzyme unit (IU) refers to:

(a) 1 mmol of substrate reacting in 1 min
(b) 1 nmol of substrate reacting in 1 min
(c) 1 mmol of substrate reacting in 1 sec
(d) 1 nmol of substrate reacting in 1 sec

52. The SI enzyme units are:

(a) King-Armstrong units
(b) Katals/Litre
(c) International Units
(d) Bodansky units

53. Katal is expressed as:

(a) Moles/sec
(b) Moles/min
(c) Mmol/sec
(d) Mmol/min

54. The following are functional plasma enzymes except:

(a) Pseudocholinesterase
(b) Lipoprotein lipase
(c) Amylase
(d) Coagulation factor VIII

55. Increased serum amylase occurs in:

(a) Acute parotitis
(b) Acute pancreatitis
(c) Both
(d) None of the above

56. The following serum analyte is a good index of recent alcohol intake:

(a) ALT (SGPT)
(b) γ-Glutamyl transferase (γ-GT)
(c) LDH-5
(d) Ceruloplasmin

57. Serum CPK-3 is increased in the following except:

(a) Myasthenia gravis
(b) Muscular dystrophy
(c) Rhabdomyolysis
(d) Myopathy

58. Total serum LDH is increased in:
(a) Megaloblastic anemia
(b) Hemolytic anemia
(c) Acute myocardial infarction
(d) All the above

59. Maximum rise in serum alkaline phosphatase (ALP) occurs in:
(a) Paget's disease
(b) Osteoporosis
(c) Osteomalacia
(d) Acute hepatitis

60. The following enzyme is used as an analytical agent:
(a) Horseradish peroxidase
(b) Glucose-6-phosphate dehydrogenase
(c) Alkaline phosphatase (ALP)
(d) All the above

61. The following enzyme is used as therapeutic agent in cystic fibrosis:
(a) Streptokinase
(b) Tissue plasminogen activator
(c) Asparaginase
(d) DNase

62. The following enzyme is used as therapeutic agent in some leukemias:
(a) Streptokinase
(b) Tissue plasminogen activator
(c) Asparaginase
(d) DNase

63. The following enzyme is used in hypodermoclysis:
(a) Asparaginase
(b) Hyaluronidase
(c) Diastase
(d) Serratiopeptidase

64. The following is not true for ribozymes:
(a) Hydrolyze phosphodiester bonds
(b) Do not obey Michaelis Menten kinetics

(c) May have a hair-pin or hammer-head like active centre
(d) Require divalent cations

65. Ribozymes may function in/as:

(a) Intron splicing (b) Peptidyl transferase
(c) Both (d) None of the above

66. The following is true for ribozymes:

(a) Less versatile and slower than protein catalysts
(b) Act only once in a chemical event and are not reused
(c) Bind to substrates mainly by H-bonding
(d) All the above

67. Abzymes are directed against:

(a) Substrate
(b) Enzyme
(c) Enzyme-substrate complex in transition state
(d) Product

68. The following is not true for abzymes:

(a) May be present naturally in blood
(b) Possess catalytic activity
(c) Do not exhibit substrate specificity
(d) Monoclonal

69. At physiological pH, the standard redox potential of a hydrogen electrode is:

(a) +0.42 V (b) –0.42 V
(c) +4.2 V (d) –4.2 V

70. The free energy of hydrolysis is >–10 kcal/mol for the following except:

(a) Phosphoenolpyruvate (b) ATP
(c) Creatine phosphate (d) 1,3-BPG

71. The free energy of hydrolysis is maximum for:

(a) Phosphoenolpyruvate (b) ATP
(c) Creatine phosphate (d) 1,3-BPG

72. The free energy of hydrolysis is least for:
(a) Phosphoenolpyruvate (b) ATP
(c) Creatine phosphate (d) 1,3-BPG

73. The free energy of hydrolysis is:
(a) More for acetyl CoA but less for ATP
(b) More for ATP but less for acetyl CoA
(c) Equal for both
(d) Not applicable for acetyl CoA

74. A reducing equivalent refers to:
(a) e^- (b) H^+
(c) $H^+ + e^-$ (d) None of the above

75. During catabolism, more reducing equivalents are extracted from:
(a) Carbohydrates (b) Lipids
(c) Proteins (d) Same for all

76. The inner mitochondrial membrane is impermeable to:
(a) Glutamate (b) Oxaloacetate
(c) Malate (d) Aspartate

77. Flavoprotein dehydrogenase is located in:
(a) Outer surface of outer mitochondrial membrane
(b) Outer surface of inner mitochondrial membrane
(c) Inner surface of inner mitochondrial membrane
(d) Mitochondrial matrix

78. The following is not true for malate aspartate shuttle:
(a) More efficient compared to glycerophosphate shuttle
(b) Limited distribution compared to glycerophosphate shuttle
(c) Does not require flavoprotein dehydrogenase
(d) Avoids the need to transport oxaloacetate across inner mitochondrial membrane

79. Both Fe^{2+} and Fe^{3+} forms of iron are critical in the functioning of:
(a) Myoglobin (b) Hemoglobin
(c) Cytochromes (d) Coenzyme Q

80. In the electron transport chain (ETC), the following reacts directly with oxygen:

(a) Coenzyme Q (b) Cytochrome b
(c) Cytochrome c (d) Cytochrome a_3

81. The cytochromes transfer the following from coenzyme Q to oxygen:

(a) H^+ (b) e^-
(c) $H^+ + e^-$ (d) None of the above

82. Cytochrome $a.a_3$ requires:

(a) Cu^{2+} (b) Mg^{2+}
(c) Mn^{2+} (d) Ca^{2+}

83. In the sequence 'NAD$^+$ to FMN to CoQ' in the ETC, the following is transferred:

(a) H^+ (b) e^-
(c) $H^+ + e^-$ (d) None of the above

84. The following is true regarding the electron-affinity of various sequential components of the ETC:

(a) Progressively decreases (b) Remains same
(c) Progressively increases (d) Varies from time to time

85. The largest protein complex in the inner mitochondrial membrane is:

(a) Complex I (b) Complex II
(c) Complex III (d) Complex IV

86. Iron sulphur proteins are absent in:

(a) Complex I (b) Complex II
(c) Complex III (d) Complex IV

87. FMN is a constituent of:

(a) Complex I (b) Complex II
(c) Complex III (d) Complex IV

88. FAD is a constituent of:

(a) Complex I (b) Complex II
(c) Complex III (d) Complex IV

89. The following inhibit complex IV by binding to Fe^{3+}-heme except:

(a) Carbon monoxide
(b) Sodium azide
(c) Potassium cyanide
(d) None of the above

90. British Anti-Lewisite (BAL) inhibits:

(a) Complex I
(b) Complex II
(c) Complex III
(d) Complex IV

91. The administration of nitrites followed by thiosulphate is an emergency treatment for poisoning by:

(a) KCN
(b) CO
(c) BAL
(d) Rotenone

92. For NAD^+-linked dehydrogenases, the P/O ratio is:

(a) 1
(b) 2
(c) 3
(d) 4

93. For FAD-linked dehydrogenases, the P/O ratio is:

(a) 1
(b) 2
(c) 3
(d) 4

94. Uncoupling protein-1 (UCP-1) is present mainly in:

(a) Brown adipose tissue
(b) White adipose tissue
(c) Liver
(d) Muscles

95. The following is not a physiological uncoupler:

(a) Free fatty acid
(b) Thyroxine
(c) Insulin
(d) Unconjugated bilirubin

96. The following is a proton ionophore:

(a) BAL
(b) KCN
(c) 2,4-Dinitrophenol
(d) Oligomycin

97. The F_0 subunit of mitochondrial F_1-F_0-ATPase is inactivated by:

(a) BAL
(b) KCN
(c) 2,4-Dinitrophenol
(d) Oligomycin

98. Compared to the mitochondrial matrix, the intermembrane space is:

(a) Alkaline (b) Neutral
(c) Acidic (d) May be any of the above

99. The following is the major source of ATP:

(a) Phosphoglycerokinase and pyruvate kinase reactions of glycolysis
(b) Creatine kinase reaction
(c) Succinate thiokinase reaction of TCA cycle
(d) Oxidative phosphorylation

100. The following is not an example of substrate-level phosphorylation:

(a) Phosphoglycerokinase and pyruvate kinase reactions of glycolysis
(b) Creatine kinase reaction
(c) Succinate thiokinase reaction of TCA cycle
(d) Oxidative phosphorylation

101. The major use of oxygen in humans is for:

(a) Oxidative phosphorylation
(b) Hydroxylation reactions
(c) Oxygenation reactions
(d) Formation of reactive oxygen species (ROS)

102. The end product of complete reduction of oxygen in humans is:

(a) H_2O_2 (b) OH^-
(c) H_2O (d) $O_2^{\cdot -}$

103. The reactive oxygen species (ROS) are implicated in the pathogenesis of:

(a) Ischemia-reperfusion injury
(b) Alzheimer's disease
(c) Diabetes mellitus
(d) All the above

104. The following physiological process(es) involve(s) reactive oxygen species (ROS):

(a) Phagocytosis
(b) Regulation of gene expression
(c) Apoptosis
(d) All the above

105. The following is not a physiological source of reactive oxygen species (ROS):

(a) Xanthine oxidase reaction
(b) Pyruvate kinase reaction
(c) Oxidative phosphorylation
(d) Peroxisomal fatty acid oxidation

106. The most reactive oxygen species (ROS) is:

(a) $O_2^{\cdot -}$
(b) $OH^{\cdot -}$
(c) H_2O_2
(d) $ONOO^-$

107. The synthesis of the following is accelerated during oxidative stress except:

(a) Malondialdehyde (MDA)
(b) Advanced lipoxidation end-products (ALE)
(c) Advanced glycation end-products (AGE)
(d) PUFA

108. The following detoxifies both H_2O_2 and lipid hydroperoxides:

(a) Superoxide dismutase (SOD)
(b) Catalase
(c) GP_x
(d) Vitamin C

109. The most potent singlet oxygen scavenger is:

(a) Vitamin A
(b) Vitamin C
(c) Vitamin E
(d) Glutathione (GSH)

110. The following is one of the 'big three antioxidants':

(a) Vitamin A
(b) Vitamin C
(c) Vitamin E
(d) Each of the above

111. The following helps to maintain the levels of vitamin E:

(a) Vitamin C (b) GSH
(c) β-Carotene (d) All the above

112. Each heme unit is derived from:

(a) 4 molecules of glycine + 4 molecules of succinyl CoA
(b) 8 molecules of glycine + 8 molecules of succinyl CoA
(c) 4 molecules of glycine + 8 molecules of succinyl CoA
(d) 8 molecules of glycine + 4 molecules of succinyl CoA

113. The following is a zinc-containing enzyme:

(a) δ-Amino levulinic acid (ALA) synthase
(b) δ-Amino levulinic acid (ALA) dehydratase
(c) Porphobilinogen (PBG) deaminase
(d) Ferrochelatase

114. Heme regulates δ-Aminolevulinic acid (ALA) synthase by:

(a) Inactivation (b) Repression
(c) Both (d) None of the above

115. δ-Aminolevulinic acid (ALA) synthase is induced by:

(a) Ethanol (b) Griseofulvin
(c) Barbiturates (d) All the above

116. δ-Aminolevulinic acid (ALA) synthase is located in:

(a) Mitochondria (b) Cytosol
(c) Nucleus (d) Lysosomes

117. In the eight-step heme biosynthetic pathway (E1 to E8), the following is mitochondrial:

(a) Coproporphyrinogen oxidase (E6)
(b) Protoporphyrinogen oxidase (E7)
(c) Ferrochelatase (E8)
(d) All the above

118. In the eight-step heme biosynthetic pathway (E1 to E8), the following is a sulphhydryl protein:

(a) ALA dehydratase (E2) (b) Ferrochelatase (E8)
(c) Both (d) None of the above

119. Soret band is detected at:

(a) 200 nm
(b) 300 nm
(c) 400 nm
(d) 500 nm

120. Inorganic lead inhibits:

(a) ALA dehydratase (E2)
(b) Ferrochelatase (E8)
(c) Both
(d) None of the above

121. Acquired porphyria is an occupational hazard in:

(a) Paint industry
(b) Battery industry
(c) Both
(d) None of the above

122. Organic lead poisoning is an occupational hazard in:

(a) Paint industry
(b) Battery industry
(c) Automobile industry
(d) All the above

123. The following enzyme is deficient in acute intermittent porphyria:

(a) ALA dehydratase (E2)
(b) Uroporphyrinogen I synthase (E3)
(c) Uroporphyrinogen decarboxylase (E5)
(d) Coproporphyrinogen oxidase (E6)

124. The following enzyme is deficient in porphyria cutanea tarda:

(a) ALA dehydratase (E2)
(b) Uroporphyrinogen I synthase (E3)
(c) Uroporphyrinogen decarboxylase (E5)
(d) Coproporphyrinogen oxidase (E6)

125. The following enzyme is deficient in variegate porphyria:

(a) ALA synthase (E1)
(b) Uroporphyrinogen I cosynthase (E4)
(c) Protoporphyrinogen oxidase (E7)
(d) Ferrochelatase (E8)

126. The following enzyme is deficient in X-linked sideroblastic anemia:

(a) ALA synthase (E1)
(b) Uroporphyrinogen I cosynthase (E4)

(c) Protoporphyrinogen oxidase (E7)
(d) Ferrochelatase (E8)

127. The following enzyme is deficient in congenital erythropoietic porphyria:

(a) ALA synthase (E1)
(b) Uroporphyrinogen I cosynthase (E4)
(c) Protoporphyrinogen oxidase (E7)
(d) Ferrochelatase (E8)

128. The following enzyme is deficient in erythropoietic protoporphyria:

(a) ALA synthase (E1)
(b) Uroporphyrinogen I cosynthase (E4)
(c) Protoporphyrinogen oxidase (E7)
(d) Ferrochelatase (E8)

129. The following enzyme is deficient in hereditary coproporphyria:

(a) ALA dehydratase (E2)
(b) Uroporphyrinogen I synthase (E3)
(c) Uroporphyrinogen decarboxylase (E5)
(d) Coproporphyrinogen oxidase (E6)

130. Porphyrias may mimick:

(a) Acute abdomen
(b) Psychiatric disorder
(c) Cutaneous disease
(d) Any of the above

131. The management of porphyria includes:

(a) Avoid ALA synthase inducers
(b) Intravenous dextrose and/or hematin
(c) β-Carotene
(d) All the above

132. Heme oxygenase system (HOS) is located in:

(a) Cytosol
(b) Mitochondria
(c) Endoplasmic reticulum
(d) Plasma membrane

133. One gram of hemoglobin yields bilirubin equal to:

(a) 35 μg (b) 35 ng
(c) 35 mg (d) 3.5 mg

134. The following is not true for Heme oxygenase system (HOS):

(a) Requires oxygen (b) Requires NADPH
(c) Produces CO (d) Repressed by heme

135. The following competes with bilirubin for binding to albumin:

(a) Free fatty acid (b) Thyroxine
(c) Sulphonamide (d) All the above

136. The following is not true for ligandin:

(a) A family of hepatic glutathione-S-transferase
(b) Cytosolic
(c) Also called protein X
(d) Bind to bilirubin

137. In the plasma, conjugated bilirubin is bound to:

(a) α_2-Globulin (b) α_1-Globulin
(c) β-Globulin (d) None of the above

138. The following is a synthetic metalloporphyrin heme analog (inhibiting heme oxygen system, HOS) with a potential to treat neonatal hyperbilirubinemia:

(a) Tin mesoporphyrin
(b) Zinc bis glycol porphyrin
(c) Chromium mesoporphyrin
(d) All the above

139. Delta bilirubin is:

(a) Unconjugated bilirubin bound to albumin
(b) Conjugated bilirubin unbound to albumin
(c) Conjugated bilirubin bound to albumin
(d) Total bilirubin

140. The P_{50} for myoglobin is:

(a) 1-5 mm Hg (b) 5-10 mm Hg
(c) 10-15 mm Hg (d) 15-20 mm Hg

141. The P_{50} for HbA_1 is:

(a) 5 mm Hg
(b) 20 mm Hg
(c) 26 mm Hg
(d) 46 mm Hg

142. The P_{50} for HbF is:

(a) 5 mm Hg
(b) 20 mm Hg
(c) 26 mm Hg
(d) 46 mm Hg

143. The oxygen-binding kinetics of myoglobin is:

(a) Sigmoid
(b) Rectangular hyperbola
(c) Linear
(d) Parabola

144. The oxygen-binding kinetics of HbA_1 is:

(a) Sigmoid
(b) Rectangular hyperbola
(c) Linear
(d) Parabola

145. The oxygen-binding kinetics of HbF is:

(a) Sigmoid
(b) Rectangular hyperbola
(c) Linear
(d) Parabola

146. The major transporter of NO in blood is:

(a) Albumin
(b) α_1-Globulin
(c) β-Globulin
(d) Hb

147. $\xi_2\varepsilon_2$ refers to:

(a) Fetal Hb
(b) Embryonic Hb
(c) Minor adult Hb
(d) Glycosylated Hb

148. $\alpha_2\gamma_2$ refers to:

(a) Fetal Hb
(b) Embryonic Hb
(c) Minor adult Hb
(d) Glycosylated Hb

149. The following is not true for methemoglobinemia:

(a) Contains Fe^{3+}-Hb
(b) Non-functional Hb
(c) May be congenital or acquired
(d) Treatment requires strong oxidizing agents

150. The mutation β6Glu→Val indicates:

(a) Thalassemia minor (b) Sickle cell anemia
(c) HbC (d) Silent Hb mutation

151. Compared to HbA_1, HbS has:

(a) Smaller negative charge (b) Higher negative charge
(c) Equal negative charge (d) No negative charge

152. The following may be asymptomatic:

(a) Sickle cell trait (b) Thalassemia minor
(c) Both (d) None of the above

153. In hydrops fetalis, the number of missing α-globin gene is/are:

(a) 1 (b) 2
(c) 3 (d) 4

154. The mutation β6Glu→Lys indicates:

(a) Thalassemia minor (b) Sickle cell anemia
(c) HbC (d) Silent Hb mutation

155. The sixth coordination position of Fe^{2+} in heme may be:

(a) Occupied by oxygen (b) Occupied by CO
(c) Vacant (d) Any of the above

156. The following is not true for HisE7 of globin chains of hemoglobin:

(a) Binds to Fe^{2+} at sixth coordination position
(b) Imparts steric hindrance hence oxygen binds to heme at an angle of 121°
(c) Decreases the affinity of CO-binding to heme
(d) Also called distal histidine

157. The number of α-helices in each globin chain of myoglobin is/are:

(a) Zero (b) 8
(c) 7 (d) 6

158. The number of α-helices in each α-globin chain of hemoglobin is/are:

(a) Zero (b) 8
(c) 7 (d) 6

159. The number of α-helices in each β-globin chain of hemoglobin is/are:

(a) Zero (b) 8
(c) 7 (d) 6

160. The strongest allosteric modulator of hemoglobin is:

(a) CO_2 (b) Cl^-
(c) 2,3-BPG (d) H^+

161. The following shifts the oxy-Hb dissociation curve to the right except:

(a) Increased 2,3-BPG (b) Alkalosis
(c) Increased CO_2 (d) Hyperthermia

162. The following has the highest oxygen-affinity:

(a) HbF (b) HbA_1
(c) HbA_{1C} (d) Myoglobin

163. Quaternary structure is absent in:

(a) HbF (b) HbA_1
(c) HbA_{1C} (d) Myoglobin

164. The number of heme prosthetic groups in each molecule of HbA_1 is/are:

(a) 1 (b) 2
(c) 3 (d) 4

165. The following amino acid is critical for proton dissociation during oxygenation of hemoglobin:

(a) His146 (b) His E7
(c) His F8 (d) None of the above

166. Acidemia may have the following effect:

(a) Stimulates oxygen release from oxy-Hb
(b) Inhibits oxygen release from oxy-Hb

(c) No effect
(d) Inactivates oxy-Hb

167. Hemoglobin binds to NO in the following state:
(a) Oxy
(b) Deoxy
(c) Ferric
(d) None of the above

168. The NO released by oxy-Hb is captured by:
(a) Albumin
(b) Glutathione
(c) Globulins
(d) None of the above

169. The following has the most flexible structure:
(a) Microtubules
(b) Intermediate filaments
(c) Microfilaments
(d) None of the above

170. The following has the maximum diameter:
(a) Microtubules
(b) Intermediate filaments
(c) Microfilaments
(d) Each has the same diameter

171. The following has an associated GTPase activity:
(a) Microtubules
(b) Intermediate filaments
(c) Microfilaments
(d) None of the above

172. The following has an associated ATPase activity:
(a) Microtubules
(b) Intermediate filaments
(c) Microfilaments
(d) None of the above

173. The following promotes microtubule disassembly:
(a) Cold temperature
(b) Increased intracellular Ca^{2+}
(c) Colchicine
(d) All the above

174. The fibrous protein plectin is associated with:
(a) Microtubules
(b) Intermediate filaments
(c) Microfilaments
(d) None of the above

175. The delivery of mRNA from the nucleus to the ribosomes requires:
(a) Microtubules
(b) Intermediate filaments
(c) Microfilaments
(d) None of the above

176. Hypermethylation of the following gene is linked to colorectal cancer:

(a) Keratin (b) Vimentin
(c) Desmin (d) Lamin

177. Microtubules form part of the locomotor machinery in:

(a) WBC (b) Fibroblasts
(c) Spermatozoa (d) All the above

178. Intermediate filaments lack:

(a) Vimentin (b) Actin
(c) Lamin (d) Desmin

179. Sarcomere is the region between:

(a) Two successive Z lines
(b) Two successive H zones
(c) Midpoint of two successive A bands
(d) Midpoint of two successive I bands

180. The following is shortened during muscle contraction:

(a) Thick filament (b) Thin filament
(c) Both (d) None of the above

181. 70-80% of adult human muscle tissue is:

(a) Proteins (b) Glycogen
(c) Lipids (d) Water

182. Muscles contain the following peptides except:

(a) Carnosine (b) Anserine
(c) Carnitine (d) β-Alanyl-3-methyl histidine

183. The assembly of monomeric G actin into polymeric F actin requires:

(a) Ca^{2+} (b) Mg^{2+}
(c) Fe^{2+} (d) Cu^{2+}

184. The following is not true for titin:

(a) Largest protein in the human body
(b) Located in extracellular matrix in muscles
(c) Also called connectin

(d) Titin gene contains largest number of introns ever found in any single gene

185. The following is an actin capping protein:

(a) Filamin (b) α-Actinin
(c) β-Actinin (d) Myosin

186. The following is a β-helical protein:

(a) Collagen (b) Tropomyosin
(c) Globin (d) Titin

187. Abnormalities in titin are associated with:

(a) Scleroderma
(b) Tibial muscular dystrophy
(c) Familial hypertrophic cardiomyopathy
(d) All the above

188. The following protein is rich in aspartate and glutamate:

(a) Troponin T (b) Troponin I
(c) Troponin C (d) Tropomyosin

189. The following is not true for Duchenne's muscular dystrophy:

(a) Increased serum creatinine in early childhood
(b) Increased serum CPK-MB
(c) Muscle bulk maintained due to replacement of muscle fibres by fibro-fatty tissue
(d) LRTI is a common cause of death

190. The following are sarcolemmal proteins except:

(a) Dystroglycan (b) Sarcoglycan
(c) Dystrophin (d) Laminin

191. The biochemical basis of fatigue is:

(a) Depletion of Ca^{2+} from the sarcoplasmic reticulum
(b) Inhibition of PFK-1 by accumulation of H^+
(c) Accumulation of lactate
(d) Glycogen depletion

192. Lohmann's reaction is catalyzed by:

(a) CPK
(b) Adenylate kinase

(c) Phosphofructokinase-1 (PFK-1)
(d) Phosphofructokinase-2 (PFK-2)

193. Myokinase is:
(a) CPK
(b) Adenylate kinase
(c) Phosphofructokinase-1 (PFK-1)
(d) Phosphofructokinase-2 (PFK-2)

194. The most fatiguable muscle fibre is:
(a) Type I
(b) Type IIA
(c) Type IIB
(d) All are equally fatiguable

195. The following are called white muscle fibres:
(a) Type I
(b) Type IIA
(c) Type IIB
(d) None of the above

196. The following are called slow-twitch muscle fibres:
(a) Type I
(b) Type IIA
(c) Type IIB
(d) None of the above

197. Normally, the major source of serum γ-glutamyl transpeptidase (γ-GT) is:
(a) Liver
(b) Kidneys
(c) Small intestine
(d) Lungs

198. Highest concentration of γ-GT normally occurs in:
(a) Liver
(b) Kidneys
(c) Small intestine
(d) Lungs

199. Increased pepsin activity in gastric juice is observed in:
(a) Zollinger-Ellison syndrome
(b) Achylia gastrica
(c) Atrophic gastritis
(d) Gastric carcinoma

200. In acute pancreatitis, the following increases in serum:
(a) Amylase
(b) Lipase
(c) Elastase-I
(d) All the above

SECTION-4

NUTRITION: MACRO AND MICRONUTRIENTS

1. One calorie is the amount of heat required to:

(a) Increase the temperature of 1 g of water from 14.5°C to 15.5°C

(b) Increase the temperature of 1 g of water from 14.5°F to 15.5°F

(c) Increase the temperature of 100 g of water from 14.5°C to 15.5°C

(d) Increase the temperature of 100 g of water from 14.5°F to 15.5°F

2. One Kcal (one Cal) is equal to:

(a) 0.418 KJ
(b) 4.18 KJ
(c) 41.8 KJ
(d) 418 KJ

3. One gram of urinary nitrogen represents the following quantity of protein:

(a) 0.625 g
(b) 6.25 g
(c) 62.5 g
(d) 625 g

4. The RQ is least for:

(a) Carbohydrates
(b) Lipids
(c) Proteins
(d) Mixed diet

5. In terms of energy content, the following are nutritionally almost equal:

(a) Carbohydrates and lipids

(b) Lipids and proteins
(c) Carbohydrates and proteins
(d) Carbohydrates, lipids and proteins

6. The caloric value of ethanol is:

(a) 0.7 Kcal/g (b) 7.0 Kcal/g
(c) 70.0 Kcal/g (d) 700.0 Kcal/g

7. The caloric value of wheat compared to that of ethanol is:

(a) Equal (b) Double
(c) Half (d) Ten times

8. The energy requirement of a lean individual compared to an obese individual of the same weight, age and gender is:

(a) Equal (b) Less
(c) More (d) Either of the above

9. Most of the basal energy requirement is used for:

(a) Active membrane transport
(b) Nerve impulse transmission
(c) Involuntary muscle contraction
(d) Thermoregulation

10. The unit of BMR is:

(a) Kcal/h (b) Kcal/m^2/h
(c) Kcal/min (d) Kcal/m^2/min

11. With each decade, the BMR in adults:

(a) Increases by 2% (b) Increases by 20%
(c) Decreases by 2% (d) Decreases by 20%

12. The following increases BMR:

(a) Caffeine (b) Nicotine
(c) Alcohol (d) All the above

13. The following increases BMR:

(a) Thyroxine (b) Corticosteroids
(c) Epinephrine (d) All the above

14. The BMR is not increased in:

(a) Addison's disease (b) Hyperthyroidism
(c) Cushing's syndrome (d) Pheochromocytoma

15. In terms of specific dynamic action (SDA), the following comparison is correct:

(a) Carbohydrates>Lipids>Proteins
(b) Proteins>Lipids>Carbohydrates
(c) Lipids>Proteins>Carbohydrates
(d) Lipids>Carbohydrates>Proteins

16. The calorie requirement while ascending stairs as compared to descending stairs is:

(a) Equal (b) Double
(c) Triple (d) Half

17. In the hypothalamus, the effect of leptin on NPY cells and POMC cells is:

(a) Stimulation of NPY and inhibition of POMC
(b) Inhibition of NPY and stimulation of POMC
(c) Stimulation of both
(d) Inhibition of both

18. The following is not true for ghrelin:

(a) Produced in the stomach and small intestine
(b) Activates NPY neurons in the arcuate nucleus
(c) Inhibits the desire for food intake
(d) Regulates short-term feeding behavior

19. The richest source (w/w) of starch is:

(a) Pulses (b) Rice
(c) Wheat (d) Potato

20. The following is not true for lignins:

(a) Non-polysaccharides
(b) Abundant in carrots
(c) An example of insoluble fibre
(d) An example of soluble fibre

21. Legumes, fruits and oats are rich in:

(a) Insoluble fibres
(b) Soluble fibres
(c) Both
(d) None of the above

22. Vegetable grains are rich in:

(a) Insoluble fibres
(b) Soluble fibres
(c) Both
(d) None of the above

23. Cholesterol content (w/w) is least in:

(a) Skimmed milk
(b) Whole milk
(c) Eggs
(d) Pulses

24. Cholesterol content (w/w) is maximum in:

(a) Skimmed milk
(b) Whole milk
(c) Eggs
(d) Pulses

25. Prolonged consumption of cholesterol-rich diet may be associated with:

(a) Cancer of breast and colon
(b) Biliary stones
(c) Atherosclerosis
(d) Any of the above

26. Prolonged consumption of cholesterol-rich diet may be associated with:

(a) Obstructive jaundice
(b) Hepatocellular jaundice
(c) Hemolytic jaundice
(d) Any of the above

27. A BMI>40 kg/m^2 is interpreted as:

(a) Normal
(b) Overweight
(c) Obesity
(d) Morbid obesity

28. A BMI of 20-25 kg/m^2 is interpreted as:

(a) Normal
(b) Overweight
(c) Obesity
(d) Morbid obesity

29. The following is a lipostat:

(a) Insulin
(b) Glucagon
(c) Leptin
(d) Ghrelin

30. Juvenile-onset obesity is mainly:

(a) Hypoplastic (b) Hyperplastic
(c) Atrophic (d) Hypertrophic

31. The following is not true for central (android) obesity:

(a) Excess intra-abdominal fat around organs
(b) Also called apple-like obesity
(c) Affects men and post-menopausal women
(d) These adipocytes are highly insulin-sensitive compared to lower body adipocytes

32. The following is not true for lower body (gynoid) obesity:

(a) Excess fat around hips and thighs
(b) Affects post-menopausal women
(c) Also called pear-like obesity
(d) Associated with less health problems compared to android obesity

33. Metabolic syndrome includes:

(a) Obesity and dyslipidemia (b) Insulin resistance
(c) Hypertension (d) All the above

34. The following is no longer recommended for the purpose of weight reduction in obesity:

(a) Thyroxine (b) Fenfluramine
(c) Sibutramine (d) All the above

35. The following is a popular prescription for the purpose of weight reduction in obesity:

(a) Pancreatic lipase inhibitor
(b) Pancreatic amylase inhibitor
(c) Salivary amylase inhibitor
(d) All the above

36. The following is not true for leptin:

(a) High gastric leptin level is linked with peptic ulcer
(b) Upregulation of hypothalamic leptin receptors is linked with obesity

(c) Subnormal serum leptin level due to mutated leptin gene is linked with obesity
(d) Subnormal serum leptin level is an orexigenic stimulus

37. Daily protein requirement is higher in:
(a) Pregnancy and lactation
(b) Convalescence
(c) Children
(d) All the above

38. The following has the best protein-sparing effect:
(a) Nucleic acids
(b) Lipids
(c) Carbohydrates
(d) Same for all

39. The % of ingested protein nitrogen retained by the body is called:
(a) Biological value
(b) Amino acid score
(c) Net protein utilization
(d) Nitrogen balance

40. The following is a pair of semi-essential amino acids:
(a) Valine, histidine
(b) Histidine, arginine
(c) Methionine, arginine
(d) Glycine, glutamate

41. The following has limiting amino acids:
(a) Maize
(b) Wheat and rice
(c) Pulses
(d) All the above

42. The limiting amino acid in wheat is:
(a) Leucine
(b) Lysine
(c) Threonine
(d) Tryptophan

43. Edema, enlarged fatty liver and hair 'flag sign' are characteristic of:
(a) Marasmus
(b) Obesity
(c) Nephrotic syndrome
(d) Kwashiorkor

44. Alcoholism is not associated with:
(a) Increased fatty acid oxidation
(b) Increased triglyceride synthesis
(c) Decreased gluconeogenesis
(d) Lactacidosis

45. In a chronic alcoholic, the hepatocyte $NADH/NAD^+$ ratio is:

(a) Unchanged (b) Decreased
(c) Increased (d) Equal to one

46. The energy provision from a balanced diet should be as follows:

(a) Carbohydrates 65%, lipids 20% and proteins 15%
(b) Carbohydrates 40%, lipids 40% and proteins 20%
(c) Carbohydrates 40%, lipids 20% and proteins 40%
(d) Carbohydrates 65%, lipids 15% and proteins 20%

47. Milk is a poor source of the following except:

(a) Iron (b) Vitamin C
(c) Essential fatty acids (d) Phosphorus

48. Active avidin is found in:

(a) Raw egg white (b) Boiled egg white
(c) Raw egg yolk (d) Boiled egg yolk

49. The following vitamin is most sensitive to high temperature and long cooking time:

(a) B_{12} (b) Thiamin (B_1)
(c) Riboflavin (B_2) (d) Niacin (B_3)

50. The following vitamin is most sensitive to microwave:

(a) B_{12} (b) Thiamin (B_1)
(c) Riboflavin (B_2) (d) Niacin (B_3)

51. The following vitamin is most sensitive to UV light:

(a) B_{12} (b) Thiamin (B_1)
(c) Riboflavin (B_2) (d) Niacin (B_3)

52. The following vitamin is most sensitive to oxidation (by atmospheric oxygen):

(a) B_{12} (b) Thiamin (B_1)
(c) Riboflavin (B_2) (d) C

53. Meat is poor in:

(a) Essential fatty acids (b) Essential amino acids
(c) Zinc (d) Iron

54. Chronic alcoholism is often associated with a deficiency of:

(a) Vitamin B_6 (b) Folic acid
(c) Vitamin B_1 (d) All the above

55. Chronic glucocorticoid therapy is often associated with a deficiency of:

(a) Calcium (b) Vitamin K
(c) Zinc (d) All the above

56. Chronic phenytoin therapy is often associated with a deficiency of:

(a) Vitamin D (b) Folic acid
(c) Vitamin K (d) All the above

57. The following is required more during pregnancy than during lactation:

(a) Vitamin C (b) Folic acid
(c) Niacin (d) Vitamin B_{12}

58. The following is required more during pregnancy than during lactation:

(a) Iron (b) Calcium
(c) Vitamin D (d) Vitamin A

59. The following is required more during lactation than during pregnancy:

(a) Iron (b) Calcium
(c) Vitamin D (d) Vitamin A

60. During pregnancy, no additional requirement is to be met for:

(a) Vitamin D (b) Calcium
(c) Phosphorus (d) Protein

61. The RDA for the following vitamin is not based on energy requirement:

(a) Thiamin (B_1) (b) Riboflavin (B_2)
(c) Niacin (B_3) (d) B_{12}

62. The commonest form of malnutrition worldwide is:

(a) Kwashiorkor (b) Marasmus
(c) Obesity (d) Beriberi

63. β-Ionine ring is present in vitamin:

(a) K (b) D
(c) E (d) A

64. The following vitamin exists in the body in 3 forms (alcohol, aldehyde and acid):

(a) B_6 (b) B_1
(c) E (d) A

65. The richest source (w/w) of carotenes is:

(a) Polar bear liver (b) Carrots
(c) Papaya (d) Mango

66. Each molecule of α, β and γ-carotene yields the following number of molecules of vitamin A, respectively:

(a) 1, 1 and 1 (b) 1, 2 and 1
(c) 2, 1 and 1 (d) 1, 1 and 2

67. The conversion of β-carotene into retinal requires the following except:

(a) β-Carotene dioxygenase (b) Bile salts
(c) Retinene reductase (d) CO_2

68. The following carry a risk of overdose except:

(a) Carotenes (b) Retinol
(c) Vitamin D (d) Vitamin K

69. The following is a plasma retinol transporting protein:

(a) Retinol binding protein (b) Transthyretin
(c) Albumin (d) All the above

70. When light falls on the retina, the first biochemical event is:

(a) Rhodopsin → All trans-retinal + Opsin
(b) All trans-retinal + Opsin → Rhodopsin
(c) All trans-retinal → All trans-retinol
(d) All trans-retinol → All trans-retinal

71. Vitamin A is stored in the liver as:

(a) All trans-retinol and 11-cis-retinol
(b) All trans-retinal and 11-cis-retinal
(c) β-Carotene
(d) None of the above

72. Maximum excitation of cyanopsin occurs by light of wavelength:

(a) 558 nm
(b) 531 nm
(c) 420 nm
(d) None of the above

73. Bitot's spots are due to a deficiency of vitamin:

(a) D
(b) A
(c) K
(d) E

74. Strictly speaking, xerophthalmia is due to a deficiency of:

(a) Retinoic acid
(b) Retinol
(c) Retinal
(d) None of the above

75. Strictly speaking, decreased dark adaptation is due to a deficiency of:

(a) Retinal/retinol
(b) Retinoic acid
(c) Opsin
(d) None of the above

76. One IU of vitamin A is equal to:

(a) 0.3 μg retinol
(b) 0.3 retinol equivalents
(c) 2 μg β-carotene
(d) All the above

77. Dermatitis may occur in the following except:

(a) Vitamin A deficiency
(b) Vitamin A toxicity
(c) Niacin deficiency
(d) Niacin toxicity

78. The following is not true for vitamin D_3:

(a) Present in tomato leaves
(b) A secosteroid
(c) Derived from 7-dehydrocholesterol
(d) Synthesis requires UV-A radiation

79. Maximum absorption of dietary vitamin D occurs in:

(a) Stomach (b) Duodenum
(c) Jejunum (d) Ileum

80. The hepatic storage form of vitamin D is:

(a) Cholecalciferol (b) 25-OH-D_3
(c) 1,25-$(OH)_2$-D_3 (d) All the above

81. 25-Hydroxylase, catalyzing the hepatic synthesis of 25-OH-D_3, is present in:

(a) Nucleus
(b) Cytosol
(c) Mitochondria
(d) Smooth endoplasmic reticulum

82. 1-α-Hydroxylase, catalyzing the renal synthesis of 1,25-$(OH)_2$-D_3, is present in:

(a) Nucleus
(b) Cytosol
(c) Mitochondria
(d) Smooth endoplasmic reticulum

83. 1-α-Hydroxylase, catalyzing the synthesis of 1,25-$(OH)_2$-D_3, is present in the following except:

(a) Liver (b) Renal tubules
(c) Bones (d) Placenta

84. 24,25-$(OH)_2$-D_3 represents:

(a) Active vitamin D_3 (b) Inactive vitamin D_3
(c) Provitamin D_3 (d) Previtamin D_3

85. The following regulates renal 1-α-hydroxylase:

(a) Serum calcitriol (b) Serum PTH
(c) Serum phosphorus (d) All the above

86. A secosteroid is:

(a) A steroid nucleus with one open ring
(b) A steroid nucleus with seven rings
(c) A steroid nucleus with all rings open
(d) A steroid nucleus with no side chains

87. Calcitriol stimulates the synthesis of:
(a) Calmodulin
(b) Osteocalcin
(c) Osteopontin
(d) All the above

88. The renal action of calcitriol is:
(a) Calcium reabsorption and phosphorus excretion
(b) Phosphorus reabsorption and calcium excretion
(c) Reabsorption of calcium and phosphorus
(d) Excretion of calcium and phosphorus

89. During hypocalcemia, calcitriol:
(a) Promotes bone mineralization
(b) Promotes bone resorption
(c) Inhibits renal calcium reabsorption
(d) Inhibits intestinal calcium absorption

90. The following is not observed in vitamin D responsive rickets:
(a) Decreased 'Ca x P' product
(b) Increased serum alkaline phosphatase
(c) Decreased urinary inorganic phosphorus
(d) Decreased serum calcitriol

91. The following is true for renal rickets:
(a) Decreased serum calcium
(b) Increased serum alkaline phosphatase
(c) Treatment with calcitriol
(d) All the above

92. Genetic deficiency of renal 1-α-hydroxylase results in:
(a) Renal rickets
(b) Vitamin D resistant rickets
(c) Vitamin D responsive rickets
(d) Hypervitaminosis D

93. The following is not true for osteomalacia:
(a) May result in pelvic bone deformities interfering with parturition
(b) May occur in prolonged bed-ridden patients

(c) Synonymous with osteoporosis
(d) Increased urinary phosphorus

94. Adequate vitamin D may be synthesized by a daily sunlight exposure for:

(a) 10 min
(b) 10 sec
(c) 10 h
(d) 10 msec

95. Bone pain may occur in:

(a) Rickets
(b) Osteoporosis
(c) Hypervitaminosis D
(d) All the above

96. The following is true for calcitriol:

(a) Regulates insulin release
(b) Helps in the action of GH
(c) Immunomodulator and anticarcinogenic
(d) All the above

97. The activation of vitamin D sequentially involves:

(a) Skin, liver and kidneys
(b) Liver, skin and kidneys
(c) Skin, kidneys and liver
(d) Kidneys, skin and liver

98. The activation of vitamin D does not involve:

(a) Liver
(b) Kidneys
(c) Pancreas
(d) Skin

99. The following has maximum biological activity:

(a) α-Tocopherol
(b) β-Tocopherol
(c) γ-Tocopherol
(d) δ-Tocopherol

100. dl-α-Tocopherol is:

(a) Natural vitamin E
(b) Synthetic vitamin E
(c) Vitamin E synthesized by intestinal flora
(d) None of the above

101. d-α-Tocopherol is:

(a) Natural vitamin E
(b) Synthetic vitamin E

(c) Vitamin E synthesized by intestinal flora
(d) None of the above

102. Decreased serum vitamin E may be associated with:
(a) Acne
(b) Abetalipoproteinemia
(c) Anemia
(d) Any of the above

103. Chain breaking antioxidant vitamin is:
(a) Vitamin D
(b) Vitamin E
(c) Vitamin B_{12}
(d) Folic acid

104. The clinical supplementation of vitamin E should be restricted to:
(a) Fibrocystic breast disease
(b) Intermittent claudication
(c) Fat malabsorption
(d) All the above

105. Menadione is:
(a) Natural vitamin K in plants
(b) Vitamin K synthesized by intestinal flora
(c) Commercially available synthetic vitamin K
(d) Inactive vitamin K

106. The following is water soluble:
(a) β-Carotene
(b) β-Tocopherol
(c) Ergocalciferol
(d) Vitamin K_3

107. 'Gla' indicates:
(a) Glacial acetic acid
(b) Modified glycine
(c) γ-Carboxy glutamate
(d) Cereal protein gliadin

108. Vitamin K is not required for the activation of coagulation factor:
(a) II
(b) VIII
(c) IX
(d) X

109. Intramuscular vitamin K_3 is administered on day-1 in newborns in a dose of:
(a) 0.1 mg
(b) 1.0 mg
(c) 10 mg
(d) 100 mg

110. The following vitamin is administered intramuscularly on day-1 in newborns:

(a) Retinol
(b) Calcitriol
(c) Menadione
(d) Thiamine

111. Warfarin inhibits:

(a) Vitamin K epoxide reductase
(b) Vitamin K carboxylase
(c) Factor VIII
(d) Fibrinogen

112. Heat labile vitamin is:

(a) Thiamin (B_1)
(b) Riboflavin (B_2)
(c) Niacin (B_3)
(d) Pyridxine (B_6)

113. Aneurine is vitamin:

(a) Thiamine (B_1)
(b) Pyridoxine (B_6)
(c) B_{12}
(d) Niacin (B_3)

114. Thiaminase is present in:

(a) Tea
(b) Coffee
(c) Shellfish
(d) All the above

115. The coenzyme form of vitamin B_1 is:

(a) Thiamine monophosphate
(b) Thiamine diphosphate
(c) Thiamine triphosphate
(d) Thiamine tetraphosphate

116. Thiamine deficiency is confirmed by:

(a) Increased RBC transketolase and increased plasma pyruvate
(b) Decreased RBC transketolase and decreased plasma pyruvate
(c) Increased RBC transketolase and decreased plasma pyruvate
(d) Decreased RBC transketolase and increased plasma pyruvate

117. Thiamine requirement is not increased in:

(a) Pregnancy
(b) Alcoholism
(c) Hypothyroidism
(d) Fever

118. Prolonged and excess consumption of raw fish results in deficiency of:

(a) Vitamin B_1
(b) Vitamin B_2
(c) Vitamin B_{12}
(d) Folic acid

119. Lactoflavin is:

(a) A synthetic flavonoid
(b) A mixture of lactate and FMN
(c) Flavoured lactose
(d) Riboflavin

120. The following chemical groups are found in FMN except:

(a) Isoalloxazine
(b) Ribitol
(c) Phosphate
(d) Adenine

121. Pyridine ring is present in:

(a) Nicotinic acid
(b) Pyridoxine
(c) Both
(d) None of the above

122. Sorghum is:

(a) Rich in leucine but poor in tryptophan
(b) Rich in tryptophan but poor in leucine
(c) Rich in both
(d) Poor in both

123. Prolonged consumption of sorghum as staple food may cause:

(a) Beriberi
(b) Pellagra
(c) Megaloblastic anemia
(d) Burning feet syndrome

124. Biosynthesis of niacin may be inhibited by:

(a) Lysine
(b) Leucine
(c) Isoleucine
(d) Tryptophan

125. In pellagra, dermatitis involves the following body part:

(a) Exposed to sunlight
(b) Protected from sunlight
(c) Soles only
(d) Palms only

126. In pellagra, neurological manifestations may be due to a deficiency of:

(a) Leucine
(b) Lysine
(c) Tryptophan
(d) Tyrosine

127. The following hypolipidemic drug causes 'niacin flush':

(a) Nicotinamide (b) Nicotinic acid
(c) Both (d) None of the above

128. One mg niacin is synthesized from the following quantity of tryptophan:

(a) 60 μg (b) 60 ng
(c) 60 g (d) 60 mg

129. Desulphhydrases, dehydratases and decarboxylases require vitamin:

(a) Thiamin (B_1) (b) Riboflavin (B_2)
(c) Pyridoxine (B_6) (d) Pantothenic acid (B_5)

130. Vitamin B_6 deficiency may be diagnosed by tryptophan loading followed by measuring urinary:

(a) Xanthurenic acid (b) Xylose
(c) Pyridoxamine (d) Dopamine

131. Fatty acid chain elongation requires vitamin:

(a) Thiamin (B_1) (b) Niacin (B_3)
(c) Pyridoxine (B_6) (d) Panthothenic acid (B_5)

132. Vitamin B_6 is used in:

(a) Hyperemesis gravidarum
(b) Post radiotherapy
(c) Supportive treatment of tuberculosis along with isoniazid
(d) All the above

133. Vitamin B_6 deficiency may occur in women using oral contraceptives (OCPs) containing:

(a) No estrogen (b) Low estrogen
(c) High estrogen (d) All the above

134. β-Alanine is present in:

(a) Vitamin A (b) Pantothenic acid
(c) Biotin (d) Vitamin C

135. Synthesis of active acetate requires:

(a) Vitamin A (b) Pantothenic acid
(c) Biotin (d) Vitamin C

136. The following condition may partly respond to pantothenic acid supplementation:

(a) Burning feet syndrome
(b) Adrenal insufficiency
(c) Muscle cramps
(d) All the above

137. Active biotin is required in reactions involving:

(a) Carboxylation
(b) Decarboxylation
(c) Desulphhydration
(d) Hydrolysis

138. Lipotropic factors include:

(a) Choline
(b) Betaine
(c) Methionine
(d) All the above

139. The following is true for lipoic acid:

(a) A vitamin-like accessory growth factor
(b) A sulphur-containing fatty acid
(c) Used in selected cases of peripheral neuropathy
(d) All the above

140. The following vitamin contains a non-essential amino acid in its structure:

(a) Folic acid
(b) Vitamin B_{12}
(c) Biotin
(d) Pantothenic acid

141. Active folate is:

(a) Folic acid
(b) Dihydrofolic acid
(c) Tetrahydrofolic acid
(d) Folinic acid

142. Folinic acid is:

(a) N^5-Formyl tetrahydrofolate
(b) N^{10}-Formyl tetrahydrofolate
(c) N^5-Methyl tetrahydrofolate
(d) N^{10}-Methyl tetrahydrofolate

143. Folate deficiency can be diagnosed by estimating urinary formiminoglutamate (FIGLU) following a test dose of:

(a) Histamine
(b) Histidine
(c) Glutamate
(d) Folinic acid

144. Folate deficiency leads to decreased synthesis of:

(a) RNA (b) DNA
(c) Both (d) None of the above

145. Folate deficiency may be associated with:

(a) Hyperhomocysteinemia (b) Neural tube defects
(c) Macrocytic anemia (d) All the above

146. The inhibitor of dihydrofolate reductase is:

(a) Methyl malonate (b) Methotrexate
(c) Methionine (d) Methylcobalamine

147. Castle's extrinsic factor is:

(a) Vitamin B_{12} (b) Folic acid
(c) Vitamin B_6 (d) Vitamin C

148. Nitrosocobalamine is:

(a) Toxic form of vitamin B_{12}
(b) Inactive vitamin B_{12}
(c) Active vitamin B_{12}
(d) Precursor of vitamin B_{12} found in plants only

149. In pernicious anemia, antibodies may be directed against:

(a) Gastric parietal cells
(b) Intrinsic factor-cobalamine complex
(c) Intrinsic factor-cobalamine ileal receptors
(d) Any of the above

150. Vitamin B_{12} activity is present in:

(a) Hydroxycobalamine (b) Nitrosocobalamine
(c) Methylcobalamine (d) All the above

151. Cobamide is:

(a) Methylcobalamine (b) Deoxyadenosylcobalamine
(c) Both (d) None of the above

152. Tissue methionine stores are maintained by vitamin:

(a) B_{12} (b) B_6
(c) B_1 (d) B_2

153. The coenzyme required for methylmalonyl CoA isomerase is:

(a) Deoxyadenosylcobalamine
(b) Methylcobalamine
(c) Nitrosocobalamine
(d) None of the above

154. Vitamin B_{12} deficiency may result in:

(a) Subacute combined degeneration of spinal cord
(b) Megaloblastic anemia
(c) Methylmalonic aciduria
(d) All the above

155. Among the B-complex vitamins, the least RDA is of:

(a) Folic acid (b) Vitamin B_{12}
(c) Vitamin B_6 (d) Vitamin B_1

156. Highest concentration of vitamin C is in:

(a) Adrenal medulla (b) Renal medulla
(c) Adrenal cortex (d) Renal cortex

157. In achlorhydria, dietary vitamin C absorption is:

(a) Increased (b) Decreased
(c) Unchanged (d) May increase or decrease

158. Hydroxylation of aromatic amino acids requires vitamin:

(a) Riboflavin (B_2) (b) Niacin (B_3)
(c) B_{12} (d) C

159. Vitamin C deficiency may result in:

(a) Macrocytic anemia (b) Microcytic anemia
(c) Normochromic anemia (d) None of the above

160. Among the water soluble vitamins, the highest RDA is of:

(a) Folic acid (b) Vitamin B_{12}
(c) Biotin (d) Vitamin C

161. In the ascorbic acid saturation test, vitamin C is measured in:

(a) WBC (b) Platelets
(c) RBC (d) Serum

162. The richest source (w/w) of vitamin C is:

(a) Amla (b) Green chillies
(c) Red chillies (d) Tomato

163. Renal oxalate stones may form following megadoses of:

(a) Vitamin A (b) Vitamin D
(c) Vitamin C (d) Folic acid

164. The following vitamin, though water soluble, may be toxic in megadoses:

(a) Vitamin B_{12} (b) Folic acid
(c) Vitamin B_6 (d) Vitamin C

165. In vitamin B_{12} and/or folate deficiency, folate is 'trapped' as:

(a) Methyl tetrahydrofolate
(b) Formyl tetrahydrofolate
(c) Methylene tetrahydrofolate
(d) Dihydrofolate

166. The following triplet consists of a macromineral, a micromineral (trace mineral) and an ultratrace mineral, respectively:

(a) Cl, Cu and Cr (b) Cu, Cr and Cl
(c) Cr, Cl and Cu (d) Cl, Cr and Cu

167. The following triplet consists of a macromineral, a micromineral (trace mineral) and an ultratrace mineral respectively:

(a) Mg, Mn and Mo (b) Mg, Mo and Mn
(c) Mn, Mg and Mo (d) Mo, Mg and Mn

168. Serum calcium exists in the following form:

(a) Ionic only
(b) Ionic, salt and protein bound
(c) Ionic and protein bound only
(d) Salt and protein bound only

169. **The net calcium content in the adult human body is:**
(a) 1.0-1.2 kg
(b) 0.10-0.12 kg
(c) 0.010-0.012 kg
(d) 10.0-12.0 kg

170. **Maximum absorption of calcium occurs in:**
(a) Stomach
(b) Duodenum
(c) Jejunum
(d) Ileum

171. **Calcium absorption is decreased by all of the following except:**
(a) Acidic pH
(b) Phosphates
(c) Oxalates
(d) Alkaline pH

172. **Hypercalciuria does not occur with:**
(a) High protein diet
(b) Restricted physical activity
(c) Thyrotoxicosis
(d) Hyperparathyroidism

173. **Hypocalcemia does not trigger the synthesis/release of:**
(a) PTH
(b) Calcitriol
(c) Calcitonin
(d) PTH and calcitonin

174. **Tetany is due to a deficiency of:**
(a) Ca
(b) Cu
(c) Cl
(d) Co

175. **Formation of bone matrix requires vitamin:**
(a) C
(b) K
(c) D
(d) All the above

176. **Tetany may occur in all of the following except:**
(a) Hypocalcemia
(b) Hypomagnesemia
(c) Acidosis
(d) Alkalosis

177. **Phosphorus exists in the body in the following form:**
(a) Inorganic
(b) Organic
(c) Both
(d) None of the above

178. The intestinal absorption of phosphate is promoted by:

(a) Alkaline phosphatase (b) Acid phosphatase
(c) Amylase (d) Aldolase

179. The normal serum 'Ca × P' product (each expressed in mg/dl) is:

(a) 10-20 (b) 20-30
(c) 30-40 (d) 40-50

180. The daily requirement of phosphorus is:

(a) Equal to that of calcium
(b) Lower than that of calcium
(c) Higher than that of calcium
(d) Negligible compared to that of calcium

181. Cutaneous vitamin D synthesis is inhibited by sunscreens with sun protection factor (SPF):

(a) 2-3 (b) 4-5
(c) 6-7 (d) ≥ 8

182. Ca^{2+} enters epithelial cells (intestine/kidneys) via:

(a) TRPV5 and TRPV6
(b) Calbindin
(c) Na^+/Ca^{2+} exchanger and Ca^{2+}-ATPase
(d) All the above

183. Ca^{2+} exits epithelial cells (intestine/kidneys) via:

(a) TRPV5 and TRPV6
(b) Calbindin
(c) Na^+/Ca^{2+} exchanger and Ca^{2+}-ATPase
(d) All the above

184. Vitamin D toxicity may occur with increased:

(a) Dietary ergocalciferol (b) Dietary cholecalciferol
(c) Calcitriol supplements (d) Sunlight exposure

185. In chronic renal failure, the following vitamin should be avoided as a supplement:

(a) A (b) D
(c) C (d) B_6

186. The following is a uremic complication:

(a) Metabolic acidosis (b) Calciphylaxis
(c) Hypoalbuminemia (d) All the above

187. In chronic kidney disease, calcium carbonate/lanthanum carbonate/sevelamer is used to:

(a) Increase intestinal phosphate absorption
(b) Decrease intestinal phosphate absorption
(c) Increase intestinal calcium absorption
(d) Decrease intestinal calcium absorption

188. The following membrane protein is required for renal tubular Mg^{2+} reabsorption:

(a) Claudin-16 (b) Claudin-19
(c) TRMP6 (d) All the above

189. Mg^{2+} is an activator of:

(a) Hexokinase (b) Acetyl CoA synthetase
(c) DNA polymerase (d) All the above

190. Magnesium toxicity may occur with:

(a) Excessive use of Mg-containing antacids
(b) Treatment with diuretics
(c) Metabolic acidosis
(d) Uremia

191. 'One way substance' refers to:

(a) Fe (b) Ca
(c) Zn (d) Mg

192. The net iron content in the adult human body is:

(a) 1 g (b) 2 g
(c) 3 g (d) 4 g

193. The following are non-heme proteins except:

(a) Transferrin (b) Ferritin
(c) Ferridoxin (d) Tryptophan pyrrolase

194. In Fe-S proteins, sulphur exists as:

(a) Sulphate (b) Sulphite
(c) Sulphide (d) Sulphone

195. Iron exists as Fe^{3+} in the following except:

(a) Transferrin (b) Ferritin

(c) Hematin (d) Nitric oxide synthase

196. In the enterocytes, paraferritin complex is located in:

(a) Brush border membrane

(b) Contraluminal membrane

(c) Basolateral membrane

(d) Tight junctions

197. The following factors promote iron absorption except:

(a) Dietary vitamin C (b) Dietary cysteine

(c) Gastric HCl (d) Dietary phytates

198. In the duodenum, mucin helps in the absorption of:

(a) Fe^{2+} (b) Fe^{3+}

(c) Heme (d) All the above

199. In the small intestine, iron is absorbed as:

(a) Fe^{2+} (b) Fe^{3+}

(c) Heme (d) All the above forms

200. Mucosal block theory explains the intestinal absorption of:

(a) Fe (b) Cu

(c) Zn (d) Ca

201. The mucosal iron receptor protein is:

(a) Apoferritin (b) Apotransferrin

(c) Ferroxidase (d) Ferridoxin

202. 'Iron response elements'(IRE) are found in:

(a) mRNA for ferritin and transferrin

(b) Intestinal secretions

(c) Blood

(d) Brush border membrane of enterocytes

203. Apotransferrin binds and transports iron as:

(a) Fe^{2+} (b) Fe^{3+}

(c) Heme (d) All the above

204. The number of iron-binding sites in each molecule of apotransferrin are:

(a) 2 (b) 4

(c) 6 (d) 8

205. Iron normally occupies the following fraction of the total iron-binding sites in transferrin:

(a) $1/3^{rd}$ (b) ½

(c) $2/3^{rd}$ (d) $4/5^{th}$

206. In iron-deficiency anemia, TIBC:

(a) Increases (b) Decreases

(c) Remains unchanged (d) Cannot be calculated

207. The following is observed during iron-overload:

(a) Increased TIBC

(b) Decreased transferrin saturation

(c) Increased tissue hemosiderin

(d) Decreased tissue ferritin

208. The following is a poor dietary source of iron:

(a) Meat (b) Egg

(c) Whole grains (d) Milk

209. Iron-deficiency anemia is unlikely in:

(a) Pregnancy (b) Hookworm infection

(c) Following gastrectomy (d) Oligomenorrhea

210. Normocytic normochromic anemia is characteristic of:

(a) Iron-deficiency (b) Vitamin B_{12} deficiency

(c) Folate deficiency (d) Recent rapid hemorrhage

211. Hemosiderosis may be observed in:

(a) Multiple blood transfusions for thalassemia major

(b) Chronic pancreatitis

(c) Alcoholic cirrhosis

(d) All the above

212. Hemochromatosis may be associated with the following except:

(a) Pancreatitis (b) Skin pigmentation
(c) Hemosiderosis (d) Shrinkage of liver

213. Of the total body copper, 50% is present in:

(a) Muscles (b) Liver
(c) Bones (d) Small intestine

214. The following is not true for ceruloplasmin:

(a) α_1-Globulin
(b) Synthesized in liver
(c) Contains 6-8 copper atoms per molecule
(d) Helps in plasma iron transport

215. The following is true for ceruloplasmin:

(a) An acute-phase reactant
(b) Also called ferroxidase I
(c) Specific copper-binding protein
(d) All the above

216. Intestinal copper absorption requires:

(a) ATP7A (b) ATP7B
(c) Ceruloplasmin (d) Ferroxidase II

217. The secretion of copper from liver into bile requires:

(a) ATP7A (b) ATP7B
(c) Ceruloplasmin (d) Ferroxidase II

218. The following is a copper-containing enzyme:

(a) Lysyl oxidase (b) Tyrosinase
(c) Dopamine β-oxidase (d) All the above

219. Cytochrome oxidase contains:

(a) Cu only (b) Fe only
(c) Cu and Fe (d) None of the above

220. Extracellular superoxide dismutase contains:

(a) Cu (b) Mg
(c) Ni (d) Mo

221. 'Kinky hair syndrome' is due to a genetic deficiency of:
(a) Keratin (b) Collagen
(c) Elastin (d) ATP7A

222. The following is not true for Wilson's disease:
(a) Decreased plasma ceruloplasmin
(b) Increased urinary copper
(c) Decreased hepatic copper
(d) Mutation in *ATP7B* gene

223. The following is not true for Wilson's disease:
(a) Kayser Fleischer rings in cornea
(b) Requires penicillin treatment
(c) Increased copper in basal ganglia
(d) Autosomal recessive

224. Of the total body iodine, 2/3rd is present in:
(a) Thyroid gland (b) Parathyroid gland
(c) Saliva (d) Semen

225. In the GIT, iodine is absorbed as:
(a) Inorganic iodide (I^-) (b) Iodine (I)
(c) Iodinium (I^+) (d) Organic iodide

226. In the plasma, iodine circulates mainly as:
(a) Inorganic iodide (I^-) (b) Iodine (I)
(c) Iodinium (I^+) (d) Organic iodide

227. Goitrogens include:
(a) Thiocyanate and perchlorate only
(b) Goitrin and isoflavones only
(c) Perchlorate and goitrin only
(d) All the above

228. The following is true for fluoride:
(a) Inhibits bacterial growth
(b) Promotes osteoporosis
(c) Promotes dental caries
(d) Deficiency results in dental fluorosis

229. Inhalation of the following is an occupational hazard and may result in Parkinsonism:

(a) Magnesium (b) Manganese
(c) Fluoride (d) Iodide

230. Intestinal absorption of Mn is reduced by the following except:

(a) Ca (b) P
(c) Ethanol (d) Fe

231. The biologically active form of Cr is:

(a) Monovalent (b) Bivalent
(c) Trivalent (d) Hexavalent

232. Glucose tolerance factor consists of:

(a) Cr^{3+}, nicotinic acid and GSH
(b) Zn^{2+}, nicotinic acid and GSH
(c) Cr^{3+}, pyridoxine and GSH
(d) Zn^{2+}, pyridoxine and GSH

233. The following is not true for Se:

(a) Component of selenoproteins
(b) Required for ubiquinone synthesis
(c) Present in expired air
(d) Absorbed mainly in the ileum

234. The following is true for Se:

(a) Component of glutathione peroxidase
(b) SeC is the 21st amino acid
(c) Selenosis is an industrial hazard
(d) All the above

235. The following is required for the release of vitamin A from hepatic stores into the blood:

(a) Se (b) Zn
(c) Cr (d) Mn

236. Gustin contains:

(a) Zn (b) Ca
(c) P (d) Se

237. The following is a Zn-containing enzyme:
(a) Lactate dehydrogenase (b) Retinene-retinal reductase
(c) Alcohol dehydrogenase (d) All the above

238. Carbonic anhydrase contains:
(a) Zn (b) Ca
(c) Mg (d) Cr

239. Insulin storage and release requires:
(a) Zn (b) Ca
(c) Fe (d) Mn

240. Zinc fingers refer to:
(a) Zinc accumulation in nail beds
(b) Structural motif in certain proteins
(c) A skin disease due to zinc deficiency
(d) Colostrum as a source of zinc

241. Metallothionein participates in the absorption of:
(a) Zn (b) Cu
(c) Both (d) None of the above

242. Vitamin B_{12} contains:
(a) Co (b) Mo
(c) Cr (d) Zn

243. Glycylglycine dipeptidase contains:
(a) Co (b) Mo
(c) Cr (d) Zn

244. Xanthine oxidase contains:
(a) Selenocysteine (b) Molybdopterin
(c) Heme (d) Biopterin

245. The following is a Mo-containing enzyme:
(a) Xanthine oxidase (b) Aldehyde oxidase
(c) Sulphite oxidase (d) All the above

246. Tubular proteinuria may occur due to toxicity of:
(a) Cadmium (b) Lead
(c) Mercury (d) Any of the above

247. Renal free water excretion should be called:

(a) Aquaresis (b) Diuresis
(c) Natriuresis (d) Dehydration

248. The most abundant nutrient present in the body is:

(a) Carbohydrate (b) Protein
(c) Lipid (d) Water

249. Water content is least in:

(a) Adipose tissue (b) Bones
(c) Muscles (d) Nervous tissue

250. Water is distributed in ICF and ECF:

(a) Equally
(b) ICF > ECF
(c) ECF > ICF
(d) Either of the above because distribution varies from time to time

251. The following is not a plasma ultrafiltrate:

(a) Interstitial fluid (b) Transcellular fluid
(c) Lymph (d) Intracellular fluid

252. Maximum water is produced by the oxidation of:

(a) Carbohydrates (b) Proteins
(c) Lipids (d) Nucleic acids

253. The major extracellular cation is:

(a) K^+ (b) Na^+
(c) Ca^{2+} (d) Mg^{2+}

254. Total body sodium content is nearly:

(a) 100 mg (b) 100 g
(c) 100 mEq (d) 100 IU

255. The stoichiometry for Na^+/K^+ pump is:

(a) 3 Na^+ in and 2 K^+ out (b) 2 K^+ in and 3 Na^+ out
(c) 3 Na^+ in and 3 K^+ out (d) 2 K^+ in and 2 Na^+ out

256. **Maximum Na^+ absorption occurs in:**
   (a) Collecting ducts
   (b) Distal convoluted tubules
   (c) Loop of Henle
   (d) Proximal convoluted tubules

257. **The following promote natriuresis except:**
   (a) Prostaglandin E_2
   (b) Prostacyclin
   (c) Brain natriuretic peptide
   (d) Aldosterone

258. **Kidney natriuretic peptide is mainly synthesized by:**
   (a) Juxtaglomerular cells
   (b) Cells of proximal convoluted tubules
   (c) Intercalated cells in cortical collecting ducts
   (d) Cells of medullary collecting ducts

259. **The following is not true for guanylin:**
   (a) Produced mainly by small intestine
   (b) A peptide hormone promoting natriuresis
   (c) Produced in response to salt ingestion
   (d) Inhibits guanylate cyclase

260. **Atrial natriuretic peptide (ANP) acts on:**
   (a) Proximal convoluted tubules
   (b) Loop of Henle
   (c) Distal convoluted tubules
   (d) Medullary collecting ducts

261. **The following is not true for ANP:**
   (a) Synthesized by atrial myocytes
   (b) A vasoconstrictor
   (c) Inhibits aldosterone secretion
   (d) Lowers blood pressure

262. **Hyponatremia despite increase in total body sodium may occur in:**
   (a) Congestive cardiac failure
   (b) Hepatic cirrhosis

(c) Nephrotic syndrome
(d) All the above

263. Hyponatremia along with decrease in total body sodium may occur in:

(a) Vomiting
(b) Diarrhea
(c) Diuretic therapy
(d) All the above

264. The following is observed in syndrome of inappropriate ADH secretion (SIADH):

(a) Hypernatremia with plasma hyperosmolality
(b) Hypernatremia with plasma hypoosmolality
(c) Hyponatremia with plasma hypoosmolality
(d) Hyponatremia with plasma hyperosmolality

265. Hyponatremia with plasma hyperosmolality occurs in:

(a) Severe hyperglycemia due to uncontrolled diabetes mellitus
(b) Hypoglycemia
(c) Syndrome of inappropriate ADH secretion (SIADH)
(d) Excessive water drinking

266. Pseudohyponatremia occurs in:

(a) Hypertriglyceridemia
(b) Hyperproteinemia
(c) Both
(d) None of the above

267. Hypernatremia is always:

(a) Hyperosmolar
(b) Hypoosmolar
(c) Isoosmolar
(d) Either of the above

268. The following is not true for Liddle syndrome:

(a) Excessive renal Na^+/water reabsorption
(b) Hyperkalemia
(c) Early onset of hypertension
(d) Treated with amiloride

269. The major intracellular cation is:

(a) Na^+
(b) Ca^{2+}
(c) Mg^{2+}
(d) K^+

270. A combination of insulin and glucose is used to treat life-threatening:

(a) Hypernatremia (b) Hyponatremia
(c) Hyperkalemia (d) Hypokalemia

271. The most abundant extracellular anion is:

(a) Cl^- (b) HCO_3^-
(c) HPO_4^{2-} (d) SO_4^{2-}

272. Highest concentration of Cl^- is in:

(a) Gastric juice (b) CSF
(c) Pancreatic juice (d) Saliva

273. Normal anion gap is:

(a) 2-6 mEq/L (b) 12-16 mEq/L
(c) 22-26 mEq/L (d) 0.2-0.6 mEq/L

274. The following is a potent activator of salivary amylase:

(a) Ca^{2+} (b) Mg^{2+}
(c) Cl^- (d) Mn^{2+}

275. Daily urinary chloride excretion is nearly:

(a) 0.5-0.8 mg (b) 5.0-8.0 mg
(c) 0.5-0.8 g (d) 5.0-8.0 g

276. Cystic fibrosis generally spares the:

(a) Brain (b) Lungs
(c) Skin (d) Pancreas

277. Cystic fibrosis affects the following ion channel:

(a) Na^+ (b) K^+
(c) Cl^- (d) HCO_3^-

278. Cystic fibrosis is characterized by:

(a) Decreased sweat Cl^- (b) Increased sweat Cl^-
(c) Normal sweat Cl^- (d) Absence of sweat Cl^-

279. In cystic fibrosis, DNase is ideally administered:

(a) As an aerosol (b) Orally
(c) Intravenously (d) Intramuscularly

280. Chloride content of gastric juice is nearly:

(a) 0.15 mmol/L (b) 1.5 mmol/L
(c) 15 mmol/L (d) 150 mmol/L

281. In Addison's disease, serum Cl^- level is generally:

(a) Increased
(b) Normal
(c) Decreased
(d) May be increased or decreased

282. Hyperchloremia may occur in:

(a) Cushing's syndrome (b) Acute renal failure
(c) Renal tubular acidosis (d) All the above

283. The following is stored in the liver:

(a) Iron (b) Vitamin B_{12}
(c) Vitamin A (d) All the above

284. The following electrolyte is maximally absorbed in the proximal convoluted tubules:

(a) Na^+ (b) K^+
(c) Mg^{2+} (d) HCO_3^-

285. The following electrolyte is least absorbed in the proximal convoluted tubules:

(a) Na^+ (b) K^+
(c) Mg^{2+} (d) HCO_3^-

286. Na^+ absorption in the nephron may occur by:

(a) Na^+/H^+ exchange (NHE3)
(b) $Na^+/K^+/2Cl^-$ cotransporter (NKCC2)
(c) Na^+/Cl^- cotransporter (NCCT)
(d) Any of the above

287. The fractional excretion of Na^+ (fraction of filtered Na^+ excreted in the urine) is normally:

(a) $<1\%$ (b) 1-10%
(c) 10-20% (d) $>20\%$

288. The fractional excretion of K^+ (fraction of filtered K^+ excreted in the urine) is normally:

(a) 1% (b) 10%
(c) 50% (d) 90%

289. 'Beer protomania' is characterized by:

(a) Euvolemia (b) Hyponatremia
(c) Hypoosmolality (d) All the above

290. Increased blood urea nitrogen with normal serum Na^+ is said to be:

(a) Hyperosmolar, isotonic
(b) Isoosmolar, isotonic
(c) Hyperosmolar, hypertonic
(d) Isoosmolar, hypertonic

291. Racecadotril inhibits the following membrane-bound enzyme to reduce water and electrolyte loss in the supportive management of diarrhea:

(a) Adenylate cyclase (b) Guanylate cyclase
(c) Na^+-K^+-ATPase (d) Enkephalinase

292. Idiogenic osmoles include:

(a) Amino acids (b) Trimethylamines
(c) Myoinositol (d) All the above

293. Congenital chloridorrhea is due to:

(a) Increased gastric HCl secretion
(b) Mutated ileal/colonic Cl^-/HCO_3^- exchanger
(c) Both
(d) None of the above

294. Bartter's syndrome is characterized by defect in renal transport of:

(a) Cl^- (b) Na^+
(c) Mg^{2+} (d) HCO_3^-

295. Normal transtubular K^+-gradient is:

(a) 1-2 (b) 2-4
(c) 4-6 (d) 6-8

296. Untreated severe hyperglycemia with osmotic diuresis is associated with:

(a) Total body K^+ deficiency (b) Normokalemia
(c) Both (d) None of the above

297. Pseudohyperkalemia occurs in:

(a) Hemolysis (b) Severe leukocytosis
(c) Severe thrombocytosis (d) All the above

298. Pseudohyperkalemia is confirmed by:

(a) Serum K^+ = Plasma K^+ (b) Serum K^+ < Plasma K^+
(c) Serum K^+ > Plasma K^+ (d) Any of the above

299. In case of life-threatening hyperkalemia, the fastest way to lower serum K^+ is:

(a) Intravenous dextrose-insulin
(b) Albuterol inhalation
(c) Intravenous loop diuretics
(d) Oral $NaHCO_3$

300. The following renal tubular defect occurs in Dent's disease (X-linked nephrolithiasis):

(a) Cl^--channel
(b) Sodium phosphate cotransporter
(c) GLUT2
(d) SGLT1

SECTION-5

DEFENCE MECHANISMS: IMMUNITY, BIOTRANSFORMATIONS, BUFFERS

1. **The following is not true for innate immunity:**
   (a) Definite lag phase following antigen challenge
   (b) Determined by the genetic make-up of an individual
   (c) No prior antigen challenge required
   (d) Present since birth

2. **The following is not true for active immunity:**
   (a) Definite lag phase following antigen challenge
   (b) No immunological memory
   (c) Not applicable in immunocompromised host
   (d) Durable

3. **Administration of antitetanus serum (ATS) is an example of:**
   (a) Natural active immunity
   (b) Artificial active immunity
   (c) Natural passive immunity
   (d) Artificial passive immunity

4. **The immunity conferred from mother to neonate via milk IgA is an example of:**
   (a) Natural active immunity
   (b) Artificial active immunity
   (c) Natural passive immunity
   (d) Artificial passive immunity

5. The immunity conferred via toxoids is an example of:

(a) Natural active immunity
(b) Artificial active immunity
(c) Natural passive immunity
(d) Artificial passive immunity

6. IFN-γ is mainly produced by:

(a) Cytotoxic T cells (b) Suppressor T cells
(c) $CD4^+$ T_H1 cells (d) $CD4^+$ T_H2 cells

7. TNF-α is mainly produced by:

(a) Cytotoxic T cells (b) Suppressor T cells
(c) $CD4^+$ T_H1 cells (d) $CD4^+$ T_H2 cells

8. IL-15 is mainly produced by:

(a) Cytotoxic T cells (b) Suppressor T cells
(c) $CD4^+$ T_H1 cells (d) $CD4^+$ T_H2 cells

9. Class I and II MHC genes are located on chromosome:

(a) 1 (b) 6
(c) 16 (d) X

10. The following MHC class represents complement proteins:

(a) I (b) II
(c) III (d) None of the above

11. The following is not true for class I MHC proteins:

(a) Expressed on all cells
(b) Present cytosolic proteins to cytotoxic T cells
(c) Assisted by CD4 proteins
(d) Lead to direct killing of target cell

12. The following HLA type is associated with increased susceptibility to myasthenia gravis:

(a) B27 (b) DR4
(c) A3 (d) B8

13. The following HLA type is associated with increased susceptibility to hemochromatosis:

(a) B27 (b) DR4
(c) A3 (d) B8

14. The following HLA type is associated with increased susceptibility to ankylosing spondylitis:

(a) B27 (b) DR4
(c) A3 (d) B8

15. The following HLA type is associated with increased susceptibility to rheumatoid arthritis:

(a) B27 (b) DR4
(c) A3 (d) B8

16. The following is/are antigen presenting cell(s) (APC):

(a) T cell and B cell (b) Macrophage
(c) Dendritic cell (d) All the above

17. The following is true for a hapten:

(a) Small molecule capable of acting as an epitope
(b) Elicits antibody response only when covalently linked to a carrier
(c) Can bind to antibody even after being detached from the carrier
(d) All the above

18. An inappropriate T_H2 response is implicated in:

(a) Urticaria (b) Food allergies
(c) Bronchial asthma (d) All the above

19. The following is not an example of anaphylaxis:

(a) Graft rejection
(b) Reaction to bee sting
(c) Reaction to penicillin
(d) Reaction to an injection of therapeutic allergen for the purpose of hyposensitization

20. Contact sensitivity to nickel is an example of hypersensitivity reaction type:

(a) I (b) II
(c) III (d) IV

21. The accumulation of immune complexes in SLE is an example of hypersensitivity reaction type:

(a) I (b) II
(c) III (d) IV

22. Myasthenia gravis is an example of hypersensitivity reaction type:

(a) I (b) II
(c) III (d) IV

23. The complement system participates in the following except:

(a) Regulation of chemotaxis
(b) Clearance of immune complexes
(c) Type I hypersensitivity
(d) Antibody response

24. The following is not true for V-region of an immunoglobulin (Ig) molecule:

(a) Determines antigen specificity
(b) No two V-regions from different persons have identical amino acid sequences
(c) Located in L-chains as well as H-chains
(d) Located in the C-terminal half

25. The following is not true for C_H region of an immunoglobulin (Ig) molecule:

(a) Determines antibody class
(b) Not involved in complement binding
(c) Responsible for crossing placenta
(d) Glycosylated

26. The following may occur as a monomer or a dimer:

(a) IgM (b) IgG
(c) IgA (d) IgE

27. The following generally occurs as a pentamer:

(a) IgM (b) IgG
(c) IgA (d) IgE

28. Complementarity determining regions (CDR) in an immunoglobulin (Ig) molecule refer to:

(a) Hypervariable sequences in the V-region
(b) C_H region
(c) J chain
(d) C_L region

29. The following enzyme cleaves an immunoglobulin (Ig) molecule into 2 Fab and 1 Fc fragments:

(a) Pepsin
(b) Papain
(c) Trypsin
(d) Chymotrypsin

30. The following enzyme cleaves an immunoglobulin (Ig) molecule into a large $F(ab')_2$ fragment:

(a) Pepsin
(b) Papain
(c) Trypsin
(d) Chymotrypsin

31. Immunoglobulin (Ig) receptors are expressed on:

(a) NK cells
(b) Mast cells
(c) Eosinophils
(d) All the above

32. The hinge region is present between:

(a) V_H and C_H1
(b) V_L and C_L
(c) C_H1 and C_H2
(d) C_H2 and C_H3

33. The following amino acid in the hinge region of an immunoglobulin (Ig) molecule facilitates flexibility and proteolytic digestion:

(a) Cysteine
(b) Histidine
(c) Proline
(d) Glycine

34. Carbohydrate content is maximum in:

(a) IgE
(b) IgD
(c) IgA
(d) IgG

35. The lowest serum concentration is of:

(a) IgE
(b) IgD
(c) IgA
(d) IgG

36. The highest serum concentration is of:

(a) IgE (b) IgD
(c) IgA (d) IgG

37. The following is most important for mucosal immunity:

(a) IgE (b) IgD
(c) IgA (d) IgG

38. The following is most important for immunity against parasites:

(a) IgE (b) IgD
(c) IgA (d) IgG

39. The following is most important for placental transfer:

(a) IgE (b) IgD
(c) IgA (d) IgG

40. A neonate can synthesize:

(a) IgG (b) IgA
(c) IgM (d) IgD

41. The following number of structural genes encode an immunoglobulin (Ig) L-chain:

(a) 1 (b) 2
(c) 3 (d) 4

42. The following number of structural genes encode an immunoglobulin (Ig) H-chain:

(a) 1 (b) 2
(c) 3 (d) 4

43. The variation in different immunoglobulin (Ig) H and L-chain classes and subclasses is called:

(a) Isotype (b) Idiotype
(c) Allotype (d) Any of the above

44. The variation in hypervariable segments of an immunoglobulin (Ig) molecule is called:

(a) Isotype (b) Idiotype
(c) Allotype (d) Any of the above

45. The variation in constant region of immunoglobulin (Ig) H-chain is called:

(a) Isotype
(b) Idiotype
(c) Allotype
(d) Any of the above

46. Monoclonal antibodies are used to:

(a) Quantify proteins
(b) Detect microorganisms
(c) Classify normal and tumor cells
(d) All the above

47. The following is not considered to be an autoimmune disease:

(a) Osteoarthritis
(b) Rheumatoid arthritis
(c) Psoriatic arthritis
(d) SLE arthritis

48. The following is an autoimmune disease:

(a) Folate deficiency anemia
(b) Pernicious anemia
(c) Iron deficiency anemia
(d) Hemorrhagic anemia

49. HIV contains:

(a) ssRNA
(b) dsRNA
(c) ssDNA
(d) dsDNA

50. HIV cannot be transmitted:

(a) Through placenta
(b) By sexual intercourse
(c) Via blood/blood products
(d) Via feco-oral route

51. The following part of HIV is directly acquired from the human host:

(a) Phospholipid bilayer
(b) p17 protein matrix
(c) p24 capsid protein
(d) dsRNA

52. HIV exhibits tropism for the following except:

(a) $CD4^+$ T lymphocytes
(b) B lymphocytes
(c) Macrophages
(d) Microglial cells

53. CCR5-D32 mutation is associated with:

(a) Increased susceptibility to HIV
(b) Decreased susceptibility to HIV

(c) Primary resistance to antiretroviral drugs
(d) Secondary resistance to antiretroviral drugs

54. HIV reverse transcriptase has the following activity:
(a) Transcription of viral RNA to cDNA
(b) Ribonuclease
(c) DNA dependent DNA polymerase
(d) All the above

55. Following is an advantage of a DNA vaccine over a conventional vaccine against HIV:
(a) Stable at high temperature
(b) Cheaper and easy to manipulate by recombinant DNA technology
(c) Induce cellular as well as humoral immunity
(d) All the above

56. Following is a disadvantage of a DNA vaccine over a conventional vaccine against HIV:
(a) Risk of integration into human DNA
(b) Immune tolerance
(c) Autoimmunity
(d) All the above

57. The following drug classes are employed under 'HAART':
(a) Reverse transcriptase inhibitors
(b) Protease inhibitors
(c) Virus entry inhibitors
(d) All the above

58. The HIV capsid protein is:
(a) p24 (b) gp120
(c) gp41 (d) p17

59. The following is an Epstein Barr virus receptor on B-cells:
(a) CD19 (b) CD20
(c) CD21 (d) CD22

60. The following leucocyte surface antigen is also called 'Fas' and 'Apo-1':

(a) CD80 (b) CD86
(c) CD95 (d) CD152

61. The following is not a 'pattern recognition receptor' (PRR):

(a) Scavenger receptor (b) Pentraxin
(c) C-type lectin (d) Protegrin

62. The following is not an antimicrobial peptide:

(a) α-Defensin (b) β-Defensin
(c) Integrin (d) Histatin

63. CD14 is rich in:

(a) Leucine (b) Lysine
(c) Isoleucine (d) Proline

64. The PRR C-reactive protein (CRP) belongs to the family of:

(a) Toll-like receptors (b) Pentraxins
(c) Integrins (d) Leucine-rich proteins

65. The ligand (pathogen-associated molecular patterns, PAMPs) for CRP is:

(a) Phosphatidylcholine (b) Lipopolysaccharide
(c) Viral carbohydrates (d) Any of the above

66. The following is not a tissue macrophage:

(a) Type B lining cell of synovium
(b) Kupffer cell
(c) Osteoclast
(d) Microglial cell

67. The following is true for nitric oxide:

(a) Produced by neutrophils and macrophages
(b) Acts on lymphocytes to enhance apoptosis
(c) Prolongs adaptive immune response
(d) All the above

68. The following is not true for Chediak-Higashi syndrome:

(a) Autosomal dominant
(b) Defective degranulation of neutrophil lysosomes
(c) Decreased NK cell responsiveness
(d) A primary immunodeficiency disease

69. The following cytokine promotes T_H2 helper T cell differentiation and proliferation:

(a) IL-1 (b) IL-2
(c) IL-3 (d) IL-4

70. The following cytokine promotes eosinophil migration and proliferation:

(a) IL-3 (b) IL-4
(c) IL-5 (d) IL-6

71. The following cytokine promotes megakaryocyte colony formation and maturation:

(a) IL-10 (b) IL-11
(c) IL-12 (d) IL-13

72. The following cytokine has opposite effects to that of IL-12:

(a) IL-17 (b) IL-21
(c) IL-23 (d) All the above

73. The following cytokine acts through type II interferon receptors:

(a) INF-α (b) IFN-β
(c) INF-γ (d) All the above

74. The following is not true for oncostatin M (OSM):

(a) A cytokine
(b) Induces synthesis of hepatic acute phase proteins
(c) Stimulates thrombopoiesis
(d) Inhibits growth of Kaposi's sarcoma cells

75. The following is true for eotaxin:

(a) Chemoattractant for eosinophils
(b) Activates eosinophils along with IL-5

(c) Antibodies to eotaxin suppress airway inflammation
(d) All the above

76. The following chemokine receptor is a co-receptor for HIV-1:

(a) CCR5 (b) CXCR4
(c) Both (d) None of the above

77. The low-affinity F_c receptors for IgG on neutrophils are:

(a) CD4 (b) CD8
(c) CD16 (d) CD32

78. The F_c receptors for IgG on eosinophils are:

(a) CD4 (b) CD8
(c) CD16 (d) CD32

79. Cytokines act in the following manner:

(a) Autocrine (b) Paracrine
(c) Endocrine (d) Any of the above

80. The N-terminal conserved structural motif of type 1 cytokine receptors is rich in:

(a) Cysteine (b) Tryptophan
(c) Serine (d) All the above

81. The C-terminal conserved structural motif of type 1 cytokine receptors contains:

(a) Trp-Ser-X-Trp-Ser (b) Tyr-Ser-X-Tyr-Ser
(c) Trp-Cys-X-Trp-Cys (d) Tyr-Cys-X-Tyr-Cys

82. The following is a ligand for the Jak-Stat signaling pathway:

(a) Erythropoietin (b) Growth hormone
(c) Cytokines (d) All the above

83. The following is true for golgins:

(a) Family of 'coiled-coil' proteins associated with golgi apparatus
(b) Involved in membrane-membrane and membrane-cytoskeleton tethering events

(c) Anti-golgin antibodies may be found in Sjogren's syndrome, rheumatoid arthritis and SLE
(d) All the above

84. The following is true for Hirata's disease:

(a) Fasting hypoglycemia and high serum insulin
(b) Autoantibodies against insulin
(c) Glutamate at position 74 in HLA-DR4 β1 chain is essential for generating polyclonal antibodies whereas alanine at the same position results in monoclonal antibodies
(d) All the above

85. The following is true for infliximab:

(a) An anti-TNF- monoclonal antibody
(b) Used to treat severe rheumatoid arthritis
(c) Used to treat selected cases of ulcerative colitis
(d) All the above

86. The following is true for etanercept:

(a) Recombinant TNF-receptor-Ig-fusion protein
(b) Inhibits TNF-α
(c) Used to treat severe rheumatoid arthritis and psoriasis
(d) All the above

87. 21-Hydroxylase gene is located in the region of HLA class:

(a) I
(b) II
(c) III
(d) None of the above

88. HLA class III includes genes for:

(a) TNF-α
(b) TNF-β
(c) Hsp70
(d) All the above

89. The following is true for fibrillarin:

(a) A 34 kd protein with rRNA 2'-o-methyltransferase activity
(b) Component of nucleolar snRNP for processing pre-rRNA
(c) Anti-fibrillarin antibodies are found in severe, progressive scleroderma
(d) All the above

90. The following is true for gephyrin:

(a) A multifunctional protein
(b) Required in molybdenum cofactor synthesis and regulation of $GABA_A$ receptors
(c) Anti-gephyrin antibodies are found in paraneoplastic stiff man syndrome
(d) All the above

91. The following is true for recoverin:

(a) A 23 kd protein present in retinal photoreceptors
(b) A calcium-binding protein regulating guanylate cyclase
(c) Anti-recoverin antibodies are found in cancer-associated retinopathy (visual paraneoplastic syndrome)
(d) All the above

92. The following autoantibodies may be found in paraneoplastic neurological syndromes (PNS):

(a) Anti-Hu (anti-neuronal nuclear antibody-1, ANNA-1)
(b) Anti-Ri (anti-neuronal nuclear antibody-2, ANNA-2)
(c) Anti-Yo (anti-Purkinje cell antibody-1, APCA-1)
(d) All the above

93. The following has a strong negative association with type 1 diabetes mellitus:

(a) HLA DR2
(b) HLA DR3
(c) HLA DR4
(d) HLA DQ8

94. The following vitamin given as a supplement may rarely act as hapten and elicit an immune reaction:

(a) Thiamine
(b) Folic acid
(c) Both
(d) None of the above

95. The following is not true for C1 inhibitor deficiency:

(a) Associated with decreased serum bradykinin
(b) May be congenital or acquired
(c) May cause angioedema
(d) Similar clinical picture may occur in some patients with normal serum C1 inhibitor receiving ACE-inhibitors

96. The best screening test for SLE is serum:

(a) ANA (b) Anti-dsDNA
(c) Anti-Sm (d) Anti-RNP

97. The most specific test for SLE that is correlated with disease activity is serum:

(a) ANA (b) Anti-dsDNA
(c) Anti-Sm (d) Anti-RNP

98. Anti-Sm antibodies found in SLE indicate antibodies against:

(a) Smooth muscle
(b) Smith antigen
(c) Smooth endoplasmic reticulum
(d) Sphingomyelin

99. Rheumatoid factor is:

(a) IgM against F_c portion of IgG
(b) IgG against F_c portion of IgG
(c) IgM against F_c portion of IgM
(d) IgG against F_c portion of IgM

100. The following marker indicates a severe and progressive course in rheumatoid arthritis:

(a) High serum rheumatoid factor
(b) Presence of serum anti-CCP
(c) Eosinophilia
(d) All the above

101. The following is true for α-fodrin:

(a) A 120 kD protein
(b) Salivary gland specific
(c) Serum anti-fodrin antibodies are found in Sjogren's syndrome
(d) All the above

102. The following serum antibodies are found in Sjogren's syndrome:

(a) Anti Ro/SS-A (b) Anti La/SS-B
(c) Both (d) None of the above

103. The major c-ANCA antigen is:

(a) Proteinase-3 (b) Myeloperoxidase
(c) Elastase (d) Cathepsin G

104. The major p-ANCA antigen is:

(a) Proteinase-3 (b) Myeloperoxidase
(c) Elastase (d) Cathepsin G

105. The following is not a p-ANCA antigen:

(a) Proteinase-3 (b) Myeloperoxidase
(c) Elastase (d) Cathepsin G

106. Low serum ACE may occur in:

(a) Sarcoidosis (b) Lymphoma
(c) Hyperthyroidism (d) Gaucher's disease

107. IgA nephropathy is characterized by:

(a) Abnormal galactosylation of IgA1
(b) Decreased IgA clearance by WBC
(c) Both
(d) None of the above

108. The following is true for anti-phospholipid syndrome:

(a) Autoantibodies against β_2-glycoprotein-1
(b) Autoantibodies against oxidized-LDL
(c) Activation of endothelial cell nuclear factor-κB (NF-κB)
(d) All the above

109. Immunosuppressive effect of mycophenolate mofetil is due to inhibition of:

(a) DNA polymerase (b) IMP dehydrogenase
(c) Xanthine oxidase (d) PRPP synthetase

110. Immunity that develops to one pathogen after a host had exposure to non-identical pathogen(s) is called:

(a) Heterologous immunity (b) Immune tolerance
(c) Autoimmunity (d) Anergy

111. 'TB gold test' measures:

(a) IL-1 (b) TNF-α
(c) IFN-γ (d) C-Reactive Protein

112. TB gold test employs antigens representing:

(a) Culture filtrate protein-10 (CFP-10)
(b) Early secretary antigenic target-6 (ESAT-6)
(c) Both
(d) None the above

113. The following is true for anergy:

(a) Involves both T and B cells
(b) Important protective mechanism against autoimmunity
(c) Regulated by 'gene related to anergy in lymphocytes' (GRAIL)
(d) All the above

114. The following is true for neutrophil gelatinase-associated lipocalin (NGAL):

(a) Important in innate immunity
(b) Plasma lipocalins transport small hydrophobic molecules, e.g. steroids, bilins, retinoids
(c) Also found in renal tubules and is an early seromarker for acute kidney injury
(d) All the above

115. Antibacterial activity of honey is due to the presence of:

(a) Sucrose (b) Lysozyme
(c) Defensin-1 (d) Fructose

116. The following are capable of recognizing and binding to the respective antigens:

(a) Monoclonal antibodies (b) Polyclonal antibodies
(c) Affibodies (d) All the above

117. Tripartite motif-containing protein 5-α (TRIM 5-α) present in blood protects against HIV in:

(a) Rhesus monkeys (b) Humans
(c) Both (d) None of the above

118. Serum anti-Epstein Barr virus nuclear antigen-1 IgG (anti-EBNA-1 IgG) is a biomarker that can predict disability and disease progression in:

(a) Myasthenia gravis (b) Graves' disease
(c) Multiple sclerosis (d) Parkinson's disease

119. Single nucleotide polymorphism in a tiny DNA segment near IL28B gene results in a differential response to pegylated interferon therapy in:

(a) Hepatitis C (b) Parkinson's disease
(c) Alzheimer's disease (d) Multiple sclerosis

120. Bhopal gas tragedy is attributed to the inhalation of:

(a) Hydrogen sulphide (b) Dichloromethane
(c) Methyl isocyanate (d) Carbon monoxide

121. During phase I xenobiotic metabolism, aromatic amines are converted to:

(a) Ketones (b) Aldehydes
(c) Acids (d) Phenols

122. During phase I xenobiotic metabolism, anilides are converted to:

(a) Ketones (b) Aldehydes
(c) Acids (d) Phenols

123. During phase I xenobiotic metabolism, aromatic hydrocarbons are converted to:

(a) Ketones (b) Aldehydes
(c) Acids (d) Phenols

124. Cytochrome P_{450} exhibits a characteristic absorption peak at 450 nm following exposure to:

(a) CO (b) CO_2
(c) O_2 (d) N_2

125. Cytochrome P_{450} participates in:

(a) Hydroxylation (b) Deamination
(c) Desulphuration (d) All the above

126. The following is not required for the action of cytochrome P_{450}:

(a) O_2
(b) NAD^+
(c) Heme
(d) Functional smooth endoplasmic reticulum

127. Cytochrome P_{450} is located in:

(a) Smooth endoplasmic reticulum
(b) Rough endoplasmic reticulum
(c) Cytosol
(d) Golgi complex

128. The following is true for cytochrome P_{448}:

(a) Present in high concentration in the lungs of smokers
(b) An isoform of cytochrome P_{450}
(c) Inactivates procarcinogens, viz. polycyclic aromatic hydrocarbons
(d) Also called aromatic hydrocarbon hydroxylases

129. Isoniazid-induced toxicity may be predicted by analyzing polymorphisms of:

(a) CYP2E1 (b) CYP2C19
(c) CYP2C9 (d) CYP2D6

130. Clopidogrel-induced toxicity may be predicted by analyzing polymorphisms of:

(a) CYP2E1 (b) CYP2C19
(c) CYP2C9 (d) CYP2D6

131. Selective serotonin reuptake inhibitor (SSRI)-induced toxicity may be predicted by analyzing polymorphisms of:

(a) CYP2C19 (b) CYP2C9
(c) CYP2D6 (d) All the above

132. The following moiety of GSH is involved in conjugation reactions to form mercapturic acid:

(a) Cysteine (b) Glutamate
(c) Glycine (d) Any of the above

133. Isoniazid is mainly metabolized through:

(a) Reduction
(b) Hydrolysis
(c) Cytochrome P_{450}-mediated hydroxylation
(d) Conjugation with acetate

134. The following biotransformation forms a more toxic product compared with the parent substrate:

(a) Prontosil → Sulphanilamide
(b) Sulphanilamide → Acetyl sulphanilamide
(c) Nicotinamide → Methyl nicotinamide
(d) All the above

135. The following is not a 'green house gas':

(a) O_3
(b) CO_2
(c) NO
(d) CH_4

136. The following is a hazard of global warming:

(a) Vector-borne infectious diseases
(b) Renal stones
(c) Heat stroke
(d) All the above

137. The following is a counter-measure against global warming:

(a) Use of CFL
(b) Use of solar heaters
(c) Recycling plastic
(d) All the above

138. The following is a correct representation of pH:

(a) $\log_{10}[H^+]$
(b) $\log_{10}[1/H^+]$
(c) $\log_e[H^+]$
(d) $\log_e[1/H^+]$

139. In the reaction $NH_4^+ \leftrightarrow NH_3 + H^+$, the conjugate acid is:

(a) NH_4^+
(b) NH_3
(c) H^+
(d) Both NH_3 and H^+

140. In the reaction $H_2PO_4^- \leftrightarrow HPO_4^{2-} + H^+$, the conjugate base is:

(a) $H_2PO_4^-$
(b) HPO_4^{2-}
(c) H^+
(d) Both HPO_4^{2-} and H^+

141. Consumption of orange juice is likely to turn the urine:

(a) Alkaline (b) Acidic

(c) Highly acidic (d) No change in urine pH

142. Phosphoric acid (H_3PO_4) is formed in the body mainly by the oxidation of:

(a) Carbohydrates (b) Lipids

(c) Nucleic acids (d) Cationic amino acids

143. Hydrochloric acid (HCl) is formed in the body mainly by the oxidation of:

(a) Carbohydrates (b) Lipids

(c) Nucleic acids (d) Cationic amino acids

144. The following dietary component is metabolized to HCO_3^- and helps to alkalinize the urine particularly in vegetarians:

(a) Organic anions (b) Organic cations

(c) Inorganic anions (d) Inorganic cations

145. The major buffer system in RBC is:

(a) Bicarbonate (b) Phosphate

(c) Hemoglobin (d) Non-heme proteins

146. The pK_a' for the phosphate buffer system, $H_2PO_4^- \leftrightarrow HPO_4^{2-} + H^+$, is:

(a) 6.8 (b) 7.4

(c) 7.8 (d) 8.6

147. The pK_a' for the bicarbonate buffer system, $H_2CO_3 \leftrightarrow HCO_3^- + H^+$, is:

(a) 6.1 (b) 6.8

(c) 7.1 (d) 7.4

148. The following is an 'open' buffer system:

(a) Bicarbonate (b) Phosphate

(c) Hemoglobin (d) Plasma proteins

149. In the plasma, generally the ratio of dissolved CO_2 to carbonic acid is:

(a) 1:400 (b) 1:40
(c) 40:1 (d) 400:1

150. The pK_a' for carbonic acid is:

(a) 3.5 (b) 4.5
(c) 2.5 (d) 5.5

151. The kidneys maintain blood pH by:

(a) Reabsorption of filtered HCO_3^-
(b) Excretion of titrable acid
(c) Excretion of NH_4^+
(d) All the above

152. The majority of titrable acid in the urine is:

(a) Inorganic phosphate ($H_2PO_4^-$)
(b) Inorganic sulphate (H_2SO_4)
(c) Inorganic chloride (HCl)
(d) Carbonic acid (H_2CO_3)

153. Total acidity of urine refers to:

(a) NH_4^+ (b) Titrable acidity
(c) NH_4^+ and titrable acidity (d) H_2CO_3

154. Acidemia indicates the following H^+ concentration in blood:

(a) >45 mmol/L (b) >45 nmol/L
(c) >45 mol/L (d) >45 pmol/L

155. Alkalemia indicates the following H^+ concentration in blood:

(a) <35 mmol/L (b) <35 nmol/L
(c) <35 mol/L (d) <35 pmol/L

156. Salicylate poisoning is associated with:

(a) Hyperventilation and respiratory alkalosis
(b) Hypoventilation and respiratory acidosis
(c) Hyperventilation and respiratory acidosis
(d) Hypoventilation and respiratory alkalosis

157. Increased renal HCO_3^- excretion is the main compensation observed in:

(a) Metabolic acidosis
(b) Metabolic alkalosis
(c) Respiratory acidosis
(d) Respiratory alkalosis

158. Increased renal H^+ excretion is the main compensation observed in:

(a) Metabolic acidosis
(b) Metabolic alkalosis
(c) Respiratory acidosis
(d) Respiratory alkalosis

159. Cardiogenic shock may be associated with:

(a) Metabolic acidosis
(b) Metabolic alkalosis
(c) Respiratory acidosis
(d) Respiratory alkalosis

160. The pK'_a for ketone bodies is:

(a) 2-3
(b) 3-4
(c) 4-5
(d) 5-6

161. Kussmaul respiration is:

(a) Rapid and deep
(b) Rapid and shallow
(c) Slow and deep
(d) Slow and shallow

162. Fruity smell of acetone in expired breath is perceived in:

(a) Uncontrolled diabetes mellitus with metabolic acidosis
(b) Salicylate poisoning with respiratory alkalosis
(c) Status asthmaticus with respiratory acidosis
(d) Prolonged gastric lavage with metabolic alkalosis

163. Increased anion gap is characteristic of:

(a) Uncontrolled diabetes mellitus with metabolic acidosis
(b) Salicylate poisoning with respiratory alkalosis
(c) Status asthmaticus with respiratory acidosis
(d) Prolonged gastric lavage with metabolic alkalosis

164. In metabolic alkalosis, compensation occurs in the following order:

(a) Hyperventilation with increased renal HCO_3^- excretion
(b) Hypoventilation with increased renal HCO_3^- excretion

(c) Hyperventilation with decreased renal HCO_3^- excretion
(d) Hypoventilation with decreased renal HCO_3^- excretion

165. In the gastric parietal cells, H^+/K^+-ATPase and Cl^-/HCO_3^- exchangers are located in:
(a) Both luminal and basolateral membranes
(b) Basolateral membrane only
(c) Luminal membrane only
(d) Luminal and basolateral membranes, respectively

166. Normally, for each molecule of HCl secreted into the gastric lumen, the anion gap in blood:
(a) Increases by 1 mmol/L (b) Decreases by 1 mmol/L
(c) Does not change (d) May increase or decrease

167. Metabolic alkalosis is generally associated with:
(a) Hyperkalemia and hypochloremia
(b) Hypokalemia and hypochloremia
(c) Hyperkalemia and hyperchloremia
(d) Hypokalemia and hyperchloremia

168. The normal base excess is:
(a) +2 to +3 mmol/L (b) –3 to -2 mmol/L
(c) –2 to +3 mmol/L (d) >3 mmol/L

169. If a blood sample is alkaline, base excess will be:
(a) Positive (b) Negative
(c) Zero (d) Positive or negative

170. The sum of all buffering agents in blood is called:
(a) Anion gap (b) Buffer base
(c) Base excess (d) Buffering capacity

171. High anion gap metabolic acidosis is seen in:
(a) Diabetic ketoacidosis (b) Lactacidosis
(c) Salicylate poisoning (d) All the above

172. High anion gap metabolic acidosis is not seen in:
(a) Uremia
(b) Methanol poisoning
(c) Paraldehyde poisoning
(d) Renal tubular acidosis (RTA)

173. In distal RTA, titrable acid excretion is:
(a) Decreased (b) Increased
(c) Unchanged (d) Zero

174. Hyperkalemia and metabolic acidosis are characteristic of:
(a) Type 1 RTA (b) Type 2 RTA
(c) Type 3 RTA (d) Type 4 RTA

175. Hypoaldosteronism may be associated with:
(a) Type 1 RTA (b) Type 2 RTA
(c) Type 3 RTA (d) Type 4 RTA

176. Lithium therapy may be associated with:
(a) Type 1 RTA (b) Type 2 RTA
(c) Type 3 RTA (d) Type 4 RTA

177. Oral acetazolamide therapy may be associated with:
(a) Type 1 RTA (b) Type 2 RTA
(c) Type 3 RTA (d) Type 4 RTA

178. The stoichiometry for Na^+/H^+ exchanger (NHE) is $Na^+ : H^+ =$:
(a) 2:1 (b) 1:2
(c) 3:2 (d) 1:1

179. H^+-ATPases in lysosomal membranes pump H^+:
(a) From lysosomes to cytosol
(b) From cytosol to lysosomes
(c) Either way
(d) From one part of lysosomal membrane to the other

180. The free acidity in gastric residue is:

(a) 0-40 meq/L (b) 40-80 meq/L
(c) 80-120 meq/L (d) 120-160 meq/L

181. The total acidity in gastric residue is:

(a) 10-50 meq/L (b) 50-90 meq/L
(c) 90-130 meq/L (d) 130-170 meq/L

182. The normal gastric basal acid output (BAO) is:

(a) 0-10 meq/h (b) 0-10 meq/sec
(c) 0-10 meq/min (d) 0-10 meq/day

183. The normal gastric maximum acid output (MAO) is:

(a) 0-35 meq/h (b) 0-35 meq/sec
(c) 0-35 meq/min (d) 0-35 meq/day

184. The most potent stimulus for gastric HCl secretion is:

(a) Gastrin (b) Histamine
(c) Insulin (d) Caffeine

185. The normal HCO_3^- content in pancreatic juice is:

(a) 0.7-1.0 meq/L (b) 7-10 meq/L
(c) 70-100 meq/L (d) 700-1000 meq/L

186. The following may cause type A lactacidosis:

(a) Metformin (b) Carbon monoxide
(c) Thiamine deficiency (d) Cyanide

187. The following binds to mitochondrial membrane complex I and may cause lactacidosis:

(a) Phenformin (b) Metformin
(c) Both (d) None of the above

188. Lactacidosis in alcoholics may be due to:

(a) Decreased hepatic conversion of lactate to glucose
(b) Increased $NADH/NAD^+$ ratio favouring conversion of pyruvate to lactate
(c) Thiamine deficiency
(d) All the above

189. D-Lactic acidosis occurs in:

(a) Short-bowel syndrome
(b) Propylene glycol poisoning
(c) Hypoglycemia
(d) Malignancy

190. Baking soda pica and metabolic alkalosis may be seen in:

(a) Iron deficiency anemia (b) Pregnancy
(c) Hypothyroidism (d) Hypervitaminosis A

191. Toluene poisoning results in formation of:

(a) Lactic acid (b) Phosphoric acid
(c) Hippuric acid (d) Citric acid

192. Cushing's syndrome may present with:

(a) Metabolic acidosis (b) Metabolic alkalosis
(c) Respiratory acidosis (d) Respiratory alkalosis

193. Commonest cause of chronic hypercapnia is:

(a) Chronic obstructive pulmonary disease
(b) Organophosphorus poisoning
(c) Rib fracture
(d) Fat embolism

194. Carbonic anhydrase inhibitors such as acetazolamide promote the urinary loss of:

(a) HCO_3^- (b) K^+
(c) Both (d) None of the above

195. The most frequently encountered acid-base imbalance is:

(a) Metabolic acidosis (b) Metabolic alkalosis
(c) Respiratory acidosis (d) Respiratory alkalosis

196. Normal pregnancy is often associated with:

(a) Metabolic acidosis (b) Metabolic alkalosis
(c) Respiratory acidosis (d) Respiratory alkalosis

197. Doxapram, a respiratory stimulant, may cause:

(a) Metabolic acidosis (b) Metabolic alkalosis
(c) Respiratory acidosis (d) Respiratory alkalosis

198. 'Delta/Delta' ratio estimates:

(a) Increase in anion gap : Increase in plasma HCO_3^-
(b) Increase in anion gap : Decrease in plasma HCO_3^-
(c) Decrease in anion gap : Increase in plasma HCO_3^-
(d) Decrease in anion gap : Decrease in plasma HCO_3^-

199. For every 1g/dl decrease in serum albumin below its normal range, anion gap:

(a) Increases by 2.5 meq/L (b) Decreases by 2.5 meq/L
(c) Increases by 0.25 meq/L (d) Decreases by 0.25 meq/L

200. Nucleoside reverse transcriptase inhibitors may cause:

(a) Type A lactacidosis (b) Type B lactacidosis
(c) Metabolic alkalosis (d) None of the above

SECTION-6

SIGNAL TRANSDUCTION, HORMONES, MOLECULAR BIOLOGY, CANCER

1. The protein connexin participates in:

(a) Primary active transport
(b) Secondary active transport
(c) Gap junctions
(d) Synaptic transmission

2. Serotonin regulates:

(a) Sleep and memory
(b) Body temperature
(c) Sensory perception
(d) All the above

3. The following neurotransmitter enhances pain perception:

(a) Substance P
(b) Enkephalins
(c) Endorphins
(d) Dynorphins

4. The following is an inhibitory neurotransmitter:

(a) Glycine
(b) GABA
(c) Both
(d) None of the above

5. The following is an excitatory neurotransmitter:

(a) Aspartate
(b) Glutamate
(c) Both
(d) None of the above

6. $GABA_A$ receptor is a:

(a) Ca^{2+}-channel
(b) K^+-channel
(c) Cl^--channel
(d) Na^+-channel

7. $GABA_B$ receptor is a:

(a) Ca^{2+}-channel
(b) K^+-channel
(c) Cl^--channel
(d) Na^+-channel

8. Increased intracellular Ca^{2+} beyond a critical level may:

(a) Activate endonucleases and phospholipases
(b) Damage the mitochondria
(c) Damage the microtubules
(d) All the above

9. β-Endorphin binds to:

(a) μ receptor
(b) δ receptor
(c) κ receptor
(d) All the above

10. Dynorphin binds to:

(a) μ receptor
(b) δ receptor
(c) κ receptor
(d) All the above

11. Enkephalins bind to:

(a) μ and δ receptors
(b) μ and κ receptors
(c) δ and κ receptors
(d) All the above

12. 'Liberins' describe:

(a) Tropic hormone release hormones
(b) Tropic hormones
(c) Ultimate hormones
(d) 2nd messengers

13. Serum cortisol peaks between:

(a) 2 AM–4 AM
(b) 4 AM–8 AM
(c) 2 PM–4 PM
(d) 4 PM–8 PM

14. Serum GH peaks during:

(a) Deep sleep
(b) Sleep onset
(c) Waking-up
(d) Fully awake

15. The following show circadian rhythm except:

(a) Thyroxin
(b) Insulin
(c) Cortisol
(d) GH

16. The receptor for the following hormone does not require tyrosine kinase activity:

(a) IGF-1 (b) GH
(c) PTH (d) PRL

17. The receptor for the following hormone does not require increased phospholipase C activity:

(a) Oxytocin
(b) PTH
(c) Calcitonin
(d) Epinephrine (via β-adrenergic receptors)

18. $G_q\alpha$ subfamily of $G\alpha$ subunit of G-proteins activate:

(a) Adenylate cyclase (b) Phospholipase C
(c) Ion channels (d) Guanylate cyclase

19. $G_0\alpha$ subfamily of G subunit of G-proteins activate:

(a) Adenylate cyclase (b) Phospholipase C
(c) Ion channels (d) Guanylate cyclase

20. A heptahelix is present in:

(a) Cytosolic receptors
(b) Nuclear receptors
(c) Tyrosine kinase-linked receptors
(d) G-protein-linked receptors

21. Caffeine is a:

(a) Selective PDE inhibitor
(b) Non-selective PDE inhibitor
(c) Selective PDE activator
(d) Non-selective PDE activator

22. The major PDE found in inflammatory and immune cells is:

(a) PDE1 (b) PDE2
(c) PDE3 (d) PDE4

23. A PDE may inactivate:

(a) ATP (b) 5′-AMP
(c) cAMP (d) Protein kinase A

24. The following is not true for serotonin:
(a) Regulates long-term memory
(b) Acts through CREB
(c) Derived from tryptophan
(d) Selective serotonin reuptake inhibitors (SSRIs) are contra-indicated in depression

25. The following is true for diacylglycerol (DAG):
(a) Obtained from PIP_2 by the action of phospholipase C
(b) A 2nd messenger
(c) Activates protein kinase C
(d) All the above

26. The following is not a 2nd messenger:
(a) Triacylglycerol
(b) Diacylglycerol
(c) IP_3
(d) Ca^{2+}

27. Calcineurin is an intracellular:
(a) Phosphatase
(b) Kinase
(c) Isomerase
(d) None of the above

28. Receptors for T_3 are located in:
(a) Cell membrane
(b) Cytosol
(c) Nucleus
(d) Mitochondria

29. Receptors for androgens are located in:
(a) Cell membrane
(b) Cytosol
(c) Nucleus
(d) Mitochondria

30. The following acts through nuclear receptors:
(a) Calcitriol
(b) Estradiol
(c) Progesterone
(d) All the above

31. Receptors for glucocorticoids are located in:
(a) Cell membrane
(b) Cytosol
(c) Nucleus
(d) Mitochondria

32. Receptors for mineralocorticoids are located in:
(a) Cell membrane
(b) Cytosol
(c) Nucleus
(d) Mitochondria

33. The following is not true for the DNA-binding region of intracellular hormone receptors:

(a) Contain Zn-fingers
(b) Rich in cysteine
(c) Rich in proline
(d) Rich in basic amino acids

34. The DNA-binding region of intracellular hormone receptors is normally blocked by:

(a) Heat shock proteins
(b) Calmodulin
(c) IP_3
(d) DAG

35. Hypothalamic PRL-inhibitory factor is:

(a) TRH
(b) CRH
(c) Dopamine
(d) ADH

36. The following is a tripeptide:

(a) ADH
(b) TRH
(c) CRH
(d) TSH

37. The following belongs to the tachykinin family:

(a) Gastrin
(b) Erythropoietin
(c) Substance P
(d) Oxytocin

38. The 'master gland of endocrine orchestra' is:

(a) Hypothalamus
(b) Pituitary gland
(c) Thyroid gland
(d) Pancreas

39. Nearly 50% of cells of anterior pituitary are:

(a) Thyrotrophs
(b) Somatotrophs
(c) Gonadotrophs
(d) Corticotrophs

40. Intrachain disulphide bonds are present in:

(a) GH
(b) Insulin
(c) PRL
(d) All the above

41. The major source of IGF-1 is:

(a) Anterior pituitary
(b) Posterior pituitary
(c) Pancreas
(d) Liver

42. IGF-1 is structurally similar to:

(a) Insulin
(b) Proinsulin
(c) C-peptide
(d) Somatostatin

43. **The following is a potent stimulus for GH release:**
   (a) Arginine
   (b) Hypoglycemia
   (c) Decreased serum free fatty acids
   (d) All the above

44. **The key determinant of fetal growth in the 3rd trimester of pregnancy is:**
   (a) Maternal GH (b) Fetal GH
   (c) Maternal insulin (d) Fetal insulin

45. **The following is an uncommon finding in acromegaly:**
   (a) Hypocalcemia
   (b) Insulin resistance
   (c) Increased BMR
   (d) Increased renal tubular reabsorption of phosphate

46. **The functions of GH include:**
   (a) Stimulates linear growth
   (b) Maintains gonadal function
   (c) Attainment of normal bone density
   (d) All the above

47. **Serum IGF-binding protein-3 (IGFBP3) is low in:**
   (a) Acromegaly (b) Dwarfism
   (c) Both (d) None of the above

48. **Treatment with recombinant GH leads to:**
   (a) Increased serum IGF-1 (b) Increased serum IGFBP3
   (c) Both (d) None of the above

49. **The aromatase activity of follicular granulosa cells is mainly controlled by:**
   (a) TSH (b) FSH
   (c) LH (d) GH

50. **Androgen production by theca cells is mainly controlled by:**
   (a) TSH (b) FSH
   (c) LH (d) GH

51. Skin pigmentation is regulated by:

(a) ACTH (b) MSH
(c) Both (d) None of the above

52. During acute alcohol intoxication, polyuria is due to:

(a) Decreased ADH release from posterior pituitary
(b) Renal failure
(c) Hyperglycemia and osmotic diuresis
(d) None of the above

53. Lithium-induced nephrogenic diabetes insipidus is due to down regulation of:

(a) V2 receptors (b) AQP2 channels
(c) Both (d) None of the above

54. The predominant activity of 17-ketosteroids is like that of:

(a) Cortisol (b) Aldosterone
(c) Androgens (d) Vitamin D

55. In the corpus luteum, the end product of steroidogenesis is:

(a) Estradiol (b) DHEA
(c) Progesterone (d) Aldosterone

56. The preferred ligand for androgen receptors is:

(a) DHEA (b) Testosterone
(c) Dihydrotestosterone (d) Estradiol

57. Chemical castration may employ:

(a) 5α-Reductase inhibitors
(b) GnRH analogs
(c) GnRH receptor antagonists
(d) All the above

58. The following is abused in some sports activities:

(a) Erythropoietin (b) Androgens
(c) Both (d) None of the above

59. The following is not true for glucocorticoids:

(a) Decrease hepatic RNA synthesis
(b) Increase gluconeogenesis

(c) Increase urinary nitrogen excretion
(d) Suppress inflammation

60. Type I and type II glucocorticoid receptors mediate the actions of:

(a) Cortisol and aldosterone, respectively
(b) Aldosterone and cortisol, respectively
(c) Cortisol only
(d) Aldosterone only

61. The following may be observed in cells actively involved in steroidogenesis:

(a) Increased cAMP
(b) Increased Ca^{2+}
(c) Increased StAR protein
(d) All the above

62. Angiotensinogen is a:

(a) β-Globulin
(b) γ-Globulin
(c) α_1-Globulin
(d) α_2-Globulin

63. The following is an octapeptide:

(a) Angiotensin III
(b) Angiotensinogen
(c) Angiotensin I
(d) Angiotensin II

64. The following is useful to treat renin-dependent hypertension:

(a) Angiotensin II receptor antagonists
(b) ACE inhibitors
(c) Diuretics
(d) All the above

65. In Cushing's syndrome, following 1 mg dexamethasone challenge at midnight, plasma cortisol at 8 AM is:

(a) 0.05-0.5 μg/dl
(b) 0.5-5.0 μg/dl
(c) >5.0 μg/dl
(d) <0.05 μg/dl

66. The following may be observed in patients with ectopic ACTH production:

(a) Hypokalemia
(b) Hypochloremia
(c) Metabolic alkalosis
(d) All the above

67. The following is not observed in Addison's disease:

(a) Hypokalemia (b) Hyponatremia
(c) Hypochloremia (d) Hypoglycemia

68. Hypoaldosteronism may be caused by:

(a) Heparin
(b) ACE inhibitors
(c) Angiotensin receptor blockers
(d) All the above

69. The following is associated with a mutation in the renal mineralocorticoid receptor:

(a) Pseudohypoaldosteronism-I (autosomal dominant)
(b) Pseudohypoaldosteronism-I (autosomal recessive)
(c) Pseudohypoaldosteronism-II (autosomal recessive, WNK1)
(d) Pseudohypoaldosteronism-II (autosomal recessive, WNK4)

70. The following is not true for thyroglobulin:

(a) Contains disulphide bonds
(b) Monomeric
(c) Glycoprotein
(d) Undergoes iodination

71. The stoichiometry for the Na^+/I^- transporter in the basal plasma membrane of thyrofollicular cells is:

(a) $Na^+ : I^- = 1 : 1$ (b) $Na^+ : I^- = 2 : 1$
(c) $Na^+ : I^- = 2 : 3$ (d) $Na^+ : I^- = 1 : 2$

72. The Na^+/I^- transporter in the basal plasma membrane of thyrofollicular cells is:

(a) Symport (b) Antiport
(c) Ionophore (d) None of the above

73. I^- crosses the luminal plasma membrane of thyrofollicular cells via:

(a) Simple diffusion (b) Na^+/I^- symport
(c) Protein pendrin (d) Na^+/I^- antiport

74. In the plasma, thyroid hormones are bound to:

(a) Thyroxine-binding globulin
(b) Transthyretin
(c) Albumin
(d) All the above

75. Thyroperoxidase (TPO) catalyzes:

(a) Iodination of thyroglobulin
(b) Coupling of iodotyrosyl residues
(c) Oxidation of I^-
(d) All the above

76. The ratio of plasma half-life of T_4:T_3 (each expressed in days) is:

(a) 1:1
(b) 1:7
(c) 7:1
(d) 1:10

77. Selenocysteine is present at the active centre of deiodinase:

(a) Type 1
(b) Type 2
(c) Type 3
(d) All the above

78. Type 1 deiodinase is present in:

(a) Liver
(b) Kidneys
(c) Thyroid gland
(d) All the above

79. Type 2 deiodinase is present in:

(a) Skeletal muscles
(b) CNS
(c) Placenta
(d) All the above

80. Type 3 deiodinase is present in:

(a) Liver
(b) Kidneys
(c) Both
(d) None of the above

81. The conversion of T_4 to rT_3 is catalyzed by deiodinase:

(a) Type 1
(b) Type 2
(c) Type 3
(d) Any of the above

82. 5′-Deiodination is less efficient:

(a) In the fetus
(b) During carbohydrate restriction

(c) In chronic illness
(d) All the above

83. Euthyroid sick syndrome is characterized by:
(a) Very low serum T_3
(b) Increased serum rT_3
(c) Mild decrease in serum T_4
(d) All the above

84. In euthyroid sick syndrome, the following is not observed:
(a) Increased serum TSH
(b) Increased serum rT_3
(c) Mild decrease in serum T_4
(d) Very low serum T_3

85. In perinatal life, thyroid hormones promote:
(a) Neuronal differentiation
(b) Development of neuronal circuits
(c) Myelination
(d) All the above

86. The following is not a known function of thyroid hormones:
(a) Increased synthesis of cytochromes in many tissues
(b) Increased synthesis of uncoupling proteins
(c) Upregulation of hepatic LDL-receptors
(d) Decreased expression of GH gene in pituitary somatotrophs

87. In myxedema of severe hypothyroidism, the myxomatous material is composed of:
(a) Hyaluronic acid
(b) Chondroitin sulphate
(c) Protein and water
(d) All the above

88. The sera of patients with Graves' disease contain:
(a) Anti-TPO antibodies
(b) Anti-thyroglobulin antibodies
(c) Anti-TSH receptor antibodies
(d) Any of the above

89. The following is not true for calcitonin:
(a) A polypeptide hormone
(b) Secreted from parafollicular C-cells of thyroid gland
(c) Actions are antagonistic to PTH
(d) Contraindicated in Paget's disease

90. The kidneys synthesize the following except:
(a) Angiotensin II (b) Renin
(c) Erythropoietin (d) Calcitriol

91. The kidneys are a source plus a target for:
(a) Angiotensin II (b) Erythropoietin
(c) Calcitriol (d) Calcitonin

92. The kidneys are a target for:
(a) Aldosterone (b) ANP
(c) ADH (d) All the above

93. Tertiary hyperparathyroidism is:
(a) Benign adenoma of one parathyroid gland, responsive to changes in serum Ca^{2+}
(b) Benign adenomas of four parathyroid glands, responsive to changes in serum Ca^{2+}
(c) Benign adenomas of four parathyroid glands, unresponsive to changes in serum Ca^{2+}
(d) Any of the above

94. The following is true for calcium-sensing receptor:
(a) Present in C-cells of thyroid and thick ascending limb of loop of Henle
(b) G-protein coupled receptor
(c) Allosterically activated by 'calcimimetics', mimicking hypercalcemia to reduce PTH secretion and treat hyperparathyroidism
(d) All the above

95. Commonest form of hereditary rickets is:
(a) X-linked hypophosphatemic rickets
(b) Autosomal dominant hypophosphatemic rickets

(c) Autosomal recessive hypophosphatemic rickets
(d) Hereditary hypophosphatemic rickets with hypercalciuria

96. Adrenomedullin has structural similarity with:
(a) Calcitonin
(b) Calcitonin gene-related peptide
(c) Amylin
(d) All the above

97. The following is true for adrenomedullin:
(a) Acts via cAMP
(b) Enhances NO production
(c) A physiologic vasodilator and diuretic
(d) All the above

98. DNA directed DNA polymerase catalyzes the following:
(a) DNA → DNA (b) DNA → RNA
(c) RNA → cDNA (d) RNA → Protein

99. During DNA replication, if the daughter duplex consists of entirely new synthesized DNA, the process is:
(a) Semi-conservative (b) Dispersive
(c) Conservative (d) None of the above

100. During DNA replication, if the parental strands are fragmented and new strands are synthesized in short segments, the process is:
(a) Semi-conservative (b) Dispersive
(c) Conservative (d) None of the above

101. The site of origin of DNA replication is rich in:
(a) AT (b) GC
(c) Any of the above (d) None of the above

102. The following is called an 'origin-binding protein':
(a) Replication protein A (b) Dna A protein
(c) DNA polymerase (d) Helicase

103. The following does not require ATP:
(a) Dna A protein (b) Helicase
(c) DNA polymerase δ (d) Replication protein A

104. Generally, topoisomerases:

(a) Remove negative supercoils or insert positive supercoils
(b) Remove positive supercoils or insert negative supercoils
(c) Any of the above
(d) None of the above

105. Reverse gyrases:

(a) Remove negative supercoils or insert positive supercoils
(b) Remove positive supercoils or insert negative supercoils
(c) Any of the above
(d) None of the above

106. The following action of topoisomerase II is blocked by camptothecin:

(a) Endonuclease (b) Ligase
(c) Both (d) None of the above

107. Topoisomerase II is inhibited by:

(a) Etoposide (b) Doxorubicin
(c) Ofloxacin (d) All the above

108. 'Linking number' refers to:

(a) Number of times a DNA strand winds around the other
(b) Number of transcription factors linked to DNA at a given time
(c) Number of H-bonds per turn of DNA double helix
(d) Number of PDE-bonds per turn of DNA double helix

109. Catenates are removed by:

(a) Topoisomerase I
(b) Topoisomerase II
(c) Topoisomerase III
(d) Reverse gyrase

110. 'Clamp-holder' function is assigned to the following subunit of DNA polymerase III:

(a) α (b) β
(c) γ (d) ε

111. The following is not involved in eukaryotic nuclear DNA synthesis:

(a) DNA polymerase α (b) DNA polymerase β
(c) DNA polymerase γ (d) DNA polymerase ε

112. Proliferation cell nuclear antigen (PCNA) associates with:

(a) Pol α (b) Pol δ
(c) Pol ε (d) Pol β

113. The following is a lagging strand replicase:

(a) Pol α (b) Pol δ
(c) Pol ε (d) Pol β

114. The following is a leading strand replicase:

(a) Pol μ (b) Pol δ
(c) Pol ε (d) Pol β

115. The 'ter' sites are rich in:

(a) AT (b) GC
(c) GT (d) AG

116. Eukaryotic Okazaki fragments contain the following number of deoxyribonucleotides:

(a) 10-20 (b) 100-200
(c) 1000-2000 (d) 10000-20000

117. On completion of DNA replication, RNA primers are:

(a) Retained unchanged (b) Partly removed
(c) Completely removed (d) Retained and methylated

118. The DNA segment between two origins of replication is called:

(a) Replicon (b) Replication fork
(c) Replisome (d) Replicase

119. The following has helicase activity in eukaryotes:

(a) Pol α (b) Pol δ
(c) Pol ε (d) Pol β

120. The following has primase activity in eukaryotes:

(a) Pol α (b) Pol δ
(c) Pol ε (d) Pol β

121. The following is a eukaryotic reverse transcriptase:

(a) DNA polymerase (b) Primase
(c) Helicase (d) Telomerase

122. Telomerase activity is minimum in:

(a) Somatic cells (b) Germ cells of gonads
(c) Cancer cells (d) All the above

123. During mtDNA synthesis, the D-loop indicates:

(a) Deoxyribonucleotide loop
(b) Dimeric loop
(c) Displacement loop
(d) Demethylated loop

124. The following is not linked to the core-DNA of nucleosomes:

(a) H1 (b) H2A
(c) H2B (d) H3

125. DNA glycosylases participate in:

(a) Base excision repair (b) Nucleotide excision repair
(c) Mismatch repair (d) SOS repair

126. The following is true for mismatch DNA repair in eukaryotes:

(a) Directed by DNA methylation
(b) Nicked strand is degraded by MutL
(c) Defective mismatch repair may result in hereditary non-polyposis colorectal cancer
(d) All the above

127. Extensive DNA damage activates 'lesion repair' as a part of:

(a) SOS repair
(b) Mismatch repair
(c) Nucleotide excision repair
(d) Base excision repair

128. An AT→GC mutation is an example of:

(a) Point mutation, transition
(b) Point mutation, transversion
(c) Frame shift mutation, insertion
(d) Frame shift mutation, deletion

129. A GC→CG mutation is an example of:

(a) Point mutation, transition
(b) Point mutation, transversion
(c) Frame shift mutation, insertion
(d) Frame shift mutation, deletion

130. 'Two-hit' hypothesis explains the molecular basis of:

(a) Von Hippel-Lindau disease
(b) Autosomal dominant polycystic kidney disease
(c) Both
(d) None of the above

131. Rifampicin binds to the following subunit of prokaryotic RNA polymerase:

(a) α (b) β
(c) β' (d) σ

132. In eukaryotes, the following enzyme is inhibited by rifampicin:

(a) RNA Polymerase I (b) RNA Polymerase II
(c) RNA Polymerase III (d) mtRNA Polymerase

133. In eukaryotes, the following enzyme is inhibited by low concentration of α-amanitin:

(a) RNA Polymerase I (b) RNA Polymerase II
(c) RNA Polymerase III (d) mtRNA Polymerase

134. In eukaryotes, the following enzyme is inhibited by high concentration of α-amanitin:

(a) RNA Polymerase I (b) RNA Polymerase II
(c) RNA Polymerase III (d) mtRNA Polymerase

135. The synthesis of nucleolar rRNA in eukaryotes requires:

(a) RNA Polymerase I (b) RNA Polymerase II
(c) RNA Polymerase III (d) All the above

136. The synthesis of mRNA in eukaryotes requires:

(a) RNA Polymerase I (b) RNA Polymerase II
(c) RNA Polymerase III (d) All the above

137. The synthesis of tRNA in eukaryotes requires:

(a) RNA Polymerase I
(b) RNA Polymerase II
(c) RNA Polymerase III
(d) All the above

138. The following is not true for ρ-factor involved in prokaryotic transcription termination:

(a) Monomeric
(b) RNA-dependent ATPase activity
(c) Primary RNA transcript has hair-pin structure
(d) Nascent RNA is temporarily bound to ρ-factor

139. The following enzyme participates in processing eukaryotic hnRNA:

(a) Guanylyl transferase
(b) Guanine-7-methyl transferase
(c) Adenylate polymerase
(d) All the above

140. The following is not true for poly(A) tail of mRNA:

(a) Stabilizes mRNA
(b) Facilitates the exit of mRNA from nucleus
(c) Synthesized by a polymerase
(d) Synthesis is template-dependent

141. The following is true for spliceosomes:

(a) Bind to hnRNA
(b) Cleave introns at 5′-site
(c) Join upstream and downstream exons
(d) All the above

142. The following is not true for nuclear ribonuclease P:

(a) A ribozyme
(b) Has exonuclease activity
(c) Trims 5′ extensions of primary tRNA transcripts
(d) Required for the transcription of various ncRNA

143. Human mitochondrial RNase P is a:

(a) Ribozyme
(b) Ribonucleoprotein
(c) Protein
(d) RNA-DNA hybrid

144. RNase H activity is associated with:

(a) RISC
(b) Dicer
(c) dsRNA
(d) ds-siRNA

145. RNA interference is not equivalent to:

(a) Post-transcriptional gene silencing
(b) Gene knock-down
(c) Gene knock-out
(d) Aptamer-based gene silencing

146. The following is not true for spiegelmers:

(a) Artificial aptamers
(b) Oligonucleotides
(c) Contain D-ribonucleotides
(d) High enzymatic stability

147. The number of amino acid codons in the genetic code is:

(a) 64
(b) 63
(c) 62
(d) 61

148. The following is not a termination codon:

(a) AUG
(b) UGA
(c) UAA
(d) UAG

149. The number of methionine-coding codons in the genetic code is:

(a) 1
(b) 2
(c) 3
(d) 4

150. The number of tryptophan-coding codons in the genetic code is:

(a) 1
(b) 2
(c) 3
(d) 4

151. The following is/are non-universal codon(s):

(a) UGA
(b) AGA
(c) AGG
(d) All the above

152. During translation, the formation of each peptide bond consumes the following number of ATP molecules:

(a) 1
(b) 2
(c) 3
(d) 4

153. The function of amino acyl-tRNA synthetase is:

(a) To activate a specific amino acid
(b) To transfer amino acid to the corresponding tRNA
(c) Deacylation of misacylated tRNA
(d) All the above

154. According to Wobble hypothesis, the degenerate position is:

(a) Base at 5′-end of anticodon
(b) First base of anticodon
(c) Third base of codon
(d) Any of the above

155. The following is not true for Shine-Dalgarno sequence:

(a) Purine-rich
(b) Nearly 10 nucleotides long
(c) Binds to 16S rRNA
(d) Conserved in eukaryotes

156. Shine-Dalgarno sequence is located:

(a) Within the coding sequence of mRNA
(b) In hnRNA and spliced during processing
(c) Upstream of initiation codon of mRNA
(d) Downstream of coding sequence of mRNA

157. In prokaryotes, the following group is added to the initiation Met-tRNA:

(a) Acetyl
(b) Formyl
(c) Methyl
(d) Carboxyl

158. The following is a translocase:

(a) Elongation factor-1
(b) Elongation factor-2
(c) Peptidyl transferase
(d) Amino acyl-tRNA synthetase

159. The following is true for peptidyl transferase:

(a) Rich in basic amino acids
(b) Rich in proline
(c) Rich in glutamate
(d) None of the above

160. The peptidyl transferase activity resides in:

(a) tRNA
(b) mRNA
(c) rRNA of large ribosomal subunit
(d) rRNA of small ribosomal subunit

161. Peptidyl transferase is involved in the following aspect of polypeptide chain synthesis:

(a) Initiation and elongation
(b) Initiation and termination
(c) Elongation and termination
(d) Elongation only

162. In prokaryotes, the following translation termination release factor has GTPase activity:

(a) Release factor-1
(b) Release factor-2
(c) Release factor-3
(d) All the above

163. Chloramphenicol inhibits:

(a) Amino acyl-tRNA synthetase
(b) Peptidyl transferase
(c) Elongation factor-1
(d) Elongation factor-2

164. Elongation factor-2 contains diphthamide, which is a modified:

(a) Histidine
(b) Lysine
(c) Glutamate
(d) Proline

165. The following inactivates rRNA except:

(a) Ricin
(b) α-Sarcin
(c) Colicin E3
(d) Fusidic acid

166. Diphtheria toxin is encoded by:
(a) *Corynebacterium diphtheriae*
(b) *Corynephage* β
(c) Mammalian respiratory epithelial cell
(d) Alveolar macrophage

167. The target for diphtheria toxin is:
(a) Amino acyl-tRNA synthetase
(b) Peptidyl transferase
(c) Elongation factor-1
(d) Elongation factor-2

168. Diphtheria toxin is:
(a) NAD^+-dependent ADP-ribosylase
(b) NAD^+-independent ADP-ribosylase
(c) Transformylase
(d) Transmethylase

169. Methylation of ε-amino group of lysine occurs in:
(a) Molecular chaperones
(b) Elongation factor-2
(c) Histones
(d) Collagens

170. Out of the total human DNA, the non-transcribable DNA occupies:
(a) 2%
(b) 28%
(c) 58%
(d) 98%

171. The non-transcribable DNA constitutes:
(a) Regulatory sequences
(b) Origin and termination of replication
(c) Centromeres and telomeres
(d) All the above

172. The following is not true for moderately repetitive DNA sequences:
(a) Do not encode proteins
(b) Part of junk DNA
(c) Do not undergo mutations
(d) Repeated nearly 350 times per genome

173. The following is true for operon except:

(a) Coordinated unit of gene expression in eukaryotes
(b) Contains structural genes
(c) Contains regulatory genes
(d) Contains control elements

174. The following is not a product of structural genes in lac operon:

(a) β-Galactosidase (b) Permease
(c) Lac repressor (d) Transacetylase

175. The following is true for leader peptide in *E. coli*:

(a) Encoded by attenuator gene of tryptophan operon
(b) Contains 2 adjacent tryptophan residues at positions 10 and 11
(c) Contains 14 amino acids
(d) All the above

176. Zinc-finger motifs are found in:

(a) Transcription factor IIIA (b) Estrogen receptors
(c) Both (d) None of the above

177. Zinc-binding sites in zinc-finger motifs have a characteristic pattern of:

(a) Histidine (b) Cysteine
(c) Both (d) None of the above

178. In leucine-zippers, leucine gets repeated at the following position:

(a) 2nd (b) 7th
(c) 12th (d) 17th

179. Histone acetylation occurs during:

(a) DNA replication (b) Transcription
(c) Both (d) None of the above

180. Group A histone acetyl transferases (HATs) are mainly involved in:

(a) DNA replication (b) Transcription
(c) Nucleosome assembly (d) None of the above

181. Group B histone acetyl transferases (HATs) are mainly involved in:

(a) DNA replication (b) Transcription
(c) Nucleosome assembly (d) None of the above

182. Following histone acetylation, the affinity of histones for DNA:

(a) Remains unchanged (b) Decreases
(c) Increases (d) Any of the above

183. Transcriptionally silent mammalian DNA is:

(a) Non-methylated (b) Partly methylated
(c) Highly methylated (d) Any of the above

184. The commonest methylated base in transcriptionally silent mammalian DNA is:

(a) Adenine (b) Thymine
(c) Guanine (d) Cytosine

185. The methyl group of 5-methyl cytosine lies in the following part of DNA double helix:

(a) Core (b) Periphery
(c) Major groove (d) Minor groove

186. The following gene is associated with the phenomenon of genetic imprinting:

(a) IGF-2 (b) KVLQT1
(c) Both (d) None of the above

187. The following is true for riboswitches:

(a) Conserved sequence in some mRNAs
(b) Located in the 5′ or 3′ untranslated region
(c) Bind specific metabolites to regulate gene expression
(d) All the above

188. The difference between apo B48 and apo B100 is due to:

(a) Genomic imprinting (b) Alternative splicing
(c) Frame shift mutation (d) RNA editing

189. The following is a mechanism to increase protein diversity without altering DNA coding sequence:

(a) Alternative splicing (b) RNA editing
(c) RNA interference (d) All the above

190. Compared to serum fibronectin, extracellular matrix fibronectin is:

(a) Larger
(b) Smaller
(c) Same size
(d) Absent in extracellular matrix

191. The following is not true for a palindrome:

(a) Symmetrical inverted repeat
(b) 4-6 nucleotides long
(c) Undergoes methylation
(d) Methylated palindrome is highly susceptible to restriction endonucleases

192. The following restriction endonucleases cleave DNA within recognition sequences:

(a) Type I (b) Type II
(c) Type III (d) Any of the above

193. Isoschizomers are:

(a) Different restriction endonucleases which cleave within identical DNA sequences
(b) Single restriction endonuclease which cleaves within non-identical DNA sequences
(c) Different restriction endonucleases which cleave within non-identical DNA sequences
(d) Single restriction endonuclease which cleaves non-specifically within identical DNA sequences

194. The following is a desired characteristic of a cloning vector:

(a) Capable of autonomous replication
(b) Contains at least 1 specific nucleotide sequence recognizable by a restriction endonuclease

(c) Carry at least 1 gene
(d) All the above

195. The small size of cloned DNA fragments is a major limitation with:

(a) Yeast artificial chromosomes
(b) Bacteriophages
(c) Plasmids
(d) Cosmids

196. When the DNA sequence is known but the location of the gene is unknown, the following is the most suitable approach to find the same:

(a) Shotgun cloning
(b) RFLP analysis
(c) SNP analysis
(d) Chromosome walking

197. PCR has applications in the following:

(a) Diagnosis of infectious diseases
(b) Mutations in cancer
(c) Forensic science
(d) All the above

198. RT-PCR stands for:

(a) Room temperature PCR
(b) Real time PCR
(c) Reverse transcriptase PCR
(d) Restricted thermal PCR

199. Compared with conventional PCR, the advantages of Q-PCR include the following except:

(a) Time saving
(b) Higher sensitivity
(c) Quantitative
(d) DNA polymerase independent

200. The following is used as a probe in recombinant DNA technology:

(a) Radiolabelled synthetic oligonucleotide
(b) Non-radiolabelled biotin-coupled oligonucleotide

(c) Antibody
(d) Any of the above

201. The following amino acid is commonly used as a matrix for attaching oligonucleotides in DNA microarray technology:

(a) Glycine (b) Histidine
(c) Lysine (d) Methionine

202. The advantage of DNA microarray technology includes:

(a) Normal and abnormal DNA sequences can be detected in parallel
(b) Different genetic tests can be run in parallel
(c) Rapid test
(d) All the above

203. The following is not true for peptide nucleic acids:

(a) Nuclease-resistant
(b) Protease-resistant
(c) Cannot be used for antisense drug therapy
(d) Analogs of DNA or RNA

204. A DNA sequence may be determined by:

(a) Southern blot (b) Northern blot
(c) Western blot (d) Any of the above

205. An RNA sequence may be determined by:

(a) Southern blot (b) Northern blot
(c) Western blot (d) Any of the above

206. A protein sequence may be determined by:

(a) Southern blot (b) Northern blot
(c) Western blot (d) Any of the above

207. The advantage of developing transgenic animals is:

(a) Analysis of tissue-specific gene expression
(b) Study results of gene expression/silencing
(c) Identify genes involved in development
(d) All the above

208. To identify the DNA sequence where a protein binds, the following is a suitable approach:

(a) DNA fingerprinting
(b) DNA footprinting
(c) PCR
(d) To develop transgenic animals

209. For mutations resulting in diseases, the following is an incorrect pair:

(a) X-chromosome, Duchenne's muscular dystrophy
(b) Chromosome 20, Adenosine deaminase deficiency
(c) Chromosome 12, Phenylketonuria
(d) Chromosome 11, α-Thalassemia

210. The following is not an X-linked disease:

(a) Myotonic dystrophy
(b) Duchenne's muscular dystrophy
(c) Becker's dystrophy
(d) Hemophilia

211. Prenatal diagnosis is possible for:

(a) Myotonic dystrophy
(b) Huntington's chorea
(c) Cystic fibrosis
(d) All the above

212. The following can be produced by recombinant DNA technology:

(a) Interferons
(b) Interleukins
(c) Monoclonal antibodies
(d) All the above

213. The principles of gene therapy include the following except:

(a) Normal gene must be identified and cloned
(b) Lethal genetic disease should be selected
(c) Germ-line cells must be employed
(d) Single gene disorders should be chosen

214. Gene augmentation means:

(a) To replace the defective gene with the normal gene
(b) To correct the specific error in the defective gene

(c) To mask the results of the defective gene
(d) Any of the above

215. The following is not an ideal disease candidate for gene therapy:

(a) Hemophilia B (b) Diabetes mellitus
(c) Lesch Nyhan syndrome (d) Phenylketonuria

216. The following enzyme deficiency may be amenable to gene therapy:

(a) Urea cycle enzymes (b) Adenosine deaminase
(c) α_1-Antitrypsin (d) All the above

217. The number of base pairs in the human genome are:

(a) 3×10^3 (b) 3×10^5
(c) 3×10^7 (d) 3×10^9

218. The 1st genome to be completely sequenced was:

(a) *Homo sapiens* (b) *Hemophilus influenzae*
(c) *Escherichia coli* (d) *Saccharomyces cerevisiae*

219. In the human genome, true genes constitute:

(a) 2-3% (b) 12-13%
(c) 20-30% (d) 95-98%

220. The following is not a characteristic of a benign tumor:

(a) Well demarcated (b) Non-invasive
(c) Metastasis (d) Well differentiated

221. High-fat low-fibre diet is associated with cancer of:

(a) Oral cavity (b) Esophagus
(c) Stomach (d) Colon

222. The following strain of *Salmonella typhimurium* is used in Ame's test:

(a) His^+ (b) His^-
(c) Arg^+ (d) Arg^-

223. An increased risk of urinary bladder cancer is an occupational hazard in workers exposed to:

(a) Azo dyes (b) Aniline
(c) Rubber (d) All the above

224. An increased risk of hepatocellular carcinoma may be due to exposure to:

(a) Butter yellow (b) Saccharin
(c) Aflatoxin B_1 (d) All the above

225. The following form of radiation is not carcinogenic:

(a) Heat (b) UV
(c) Radionuclides (d) X-rays

226. Epstein Barr virus (EBV) is associated with:

(a) Nasopharyngeal carcinoma
(b) Burkitt's lymphoma
(c) B cell lymphoma
(d) All the above

227. The following is an RNA-containing oncogenic virus:

(a) EBV (b) HBV
(c) HTLL-1 (d) HPV

228. In some B cell lymphomas, a chromosomal translocation occurs between:

(a) 8 and 14 (b) 4 and 18
(c) 8 and 16 (d) 6 and 18

229. Cancer of the following origin is known to carry 'ras mutations':

(a) Pancreas (b) Thyroid
(c) Lung (d) All the above

230. The following oncoprotein is associated with astrocytoma:

(a) GTP-binding protein (b) Tyrosine kinase
(c) PDGF-β (d) EGF receptor

231. The 'guardian of the genome' is:

(a) pRb (b) p53
(c) c-erb B1 (d) c-erb B2

232. The following is an anti-oncogene:

(a) pRb (b) hst-1
(c) c-erb B1 (d) c-erb B2

233. The following is not an anti-oncogene:

(a) BRCA-1 (b) NF-1
(c) DCC (d) int-2

234. Hereditary non-polyposis colon cancer (HNCC) is mainly attributed to a defect in genes regulating:

(a) DNA repair (b) Apoptosis
(c) Nuclear transcription (d) Cell cycle check points

235. In a tumor cell, the following are generally increased except:

(a) Ribonucleotide reductase activity
(b) RNA synthesis
(c) Pyrimidine catabolism
(d) Rate of glycolysis

236. A tumor marker is a substance:

(a) Produced by a tumor and secreted into the blood
(b) Produced by a tumor but confined within it
(c) Produced by the normal host cells in response to a tumor
(d) Any of the above

237. The following is not a chemopreventive agent:

(a) Diallyl sulphide (b) Flavonoids
(c) Ascorbic acid (d) Phorbol ester

238. Serum CEA is a marker for the following tumors except:

(a) Colorectal (b) Prostate
(c) Lungs (d) Breast

239. Serum AFP is a tumor marker for:

(a) Ovary (b) Testis
(c) Liver (d) Tongue

240. Serum CA19-9 is a tumor marker for:

(a) Ovary (b) Testis
(c) Liver (d) Pancreas

241. Serum CA125 is a tumor marker for:

(a) Ovary (b) Testis
(c) Liver (d) Tongue

242. Serum CA15.3 is a tumor marker for:

(a) Ovary (b) Testis
(c) Liver (d) Pancreas

243. The following may be observed in multiple myeloma except:

(a) Increased urinary hydroxyproline
(b) Increased urinary Ig light chain
(c) Increased serum β2-microglobulin
(d) Decreased CSF polyamines

244. The critical event in the extrinsic apoptotic pathway is:

(a) Release of cytochrome c from mitochondria into cytosol
(b) Activation of procaspase-8
(c) Activation of procaspase-9
(d) Formation of apoptosome

245. The critical event in the intrinsic apoptotic pathway is:

(a) Release of cytochrome c from mitochondria into cytosol
(b) Activation of procaspase-8
(c) Activation of procaspase-9
(d) Formation of apoptosome

246. Excessive apoptosis has been linked with:

(a) Amyotrophic lateral sclerosis
(b) Alzheimer's disease
(c) Huntington's disease
(d) All the above

247. The following is not true for threonine aspartase-1 (taspase-1):

(a) An endopeptidase
(b) Cleaves substrates before aspartate residues
(c) Regulates TFIIA
(d) Increased expression of taspase-1 is found in many cancers

248. Programmed cell death resulting specifically from a loss of cell membrane-ECM interactions is called:

(a) Anoikis
(b) Necrosis
(c) Degeneration
(d) De-differentiation

249. Palifermin is a recombinant human:

(a) Fibroblast growth factor
(b) Epidermal growth factor
(c) Keratinocyte growth factor
(d) Nerve growth factor

250. To prevent/treat cancer chemotherapy-induced mucositis, the following is likely to be useful:

(a) Fibroblast growth factor
(b) Epidermal growth factor
(c) Keratinocyte growth factor
(d) Nerve growth factor

251. Nuclear factor kappa-B (NF-κB) regulates genes involved in:

(a) Apoptosis
(b) Inflammation and immunity
(c) Carcinogenesis
(d) All the above

252. The inactive form of NF-κB is found in the:

(a) Cytoplasm
(b) Nucleoplasm
(c) Nucleolus
(d) Chromatin

253. The following interleukin is the most potent activator of NF-κB:

(a) IL-1
(b) IL-2
(c) IL-13
(d) IL-16

254. An isochromosome consists of:

(a) 2 short (p) arms
(b) 2 long (q) arms
(c) Either of the above
(d) None of the above

255. The commonest isochromosome found in cancer cells is:

(a) 17q
(b) 17p
(c) 18q
(d) 18p

256. Trisomy 14 (Edward's syndrome) may be associated with isochromosome:

(a) 17q (b) 17p
(c) 18q (d) 18p

257. The class of proteins called ephrins participate in:

(a) Axon growth (b) Cell migration
(c) Angiogenesis (d) All the above

258. Thyroid transcription factor-1 (TTF-1) is most likely to be detected in adenocarcinoma of:

(a) Breast (b) Gastrointestinal tract
(c) Lungs (d) Genitourinary tract

259. Normally, TTF-1 is least expressed in:

(a) Type I pneumocytes
(b) Type II pneumocytes
(c) Thyrofollicular cells
(d) Thyroid parafollicular C-cells

260. The following is not true for TTF-1:

(a) A 38 kD cytosolic protein
(b) Regulates α-MSH and agouti-related peptide (AgRP) and thus controls feeding behavior in hypothalamus
(c) Regulates the transcription of thyroglobulin and thyroperoxidase
(d) Regulates the transcription of surfactant and Clara cell secretory protein

261. The following are tumor suppressor genes:

(a) Gatekeeper genes (b) Caretaker genes
(c) Landscaper genes (d) All the above

262. 'Oncogene addiction' is observed in:

(a) HER2 (breast cancer)
(b) ABL (Chronic myelocytic leukemia)
(c) VEGF (tumor angiogenesis)
(d) All the above

263. To prevent cancer chemotherapy-induced emesis, the following is useful:

(a) Dopamine receptor antagonists
(b) Serotonin receptor antagonists
(c) Substance P (neurokinin-1) receptor antagonists
(d) All the above

264. For cancer-related bone pain, the following is useful:

(a) Strontium-89 (b) Samarium-153
(c) Both (d) None of the above

265. The following is not an epigenetic change:

(a) Histone acetylation (b) Histone methylation
(c) Frame shift mutation (d) Methylation of CpG islands

266. The following is not true for CpG islands:

(a) Present in transcription regulatory region of active genes
(b) Occur in high frequency
(c) Normally exist in methylated state
(d) Site of epigenetic change

267. P-glycoprotein spans the plasma membrane:

(a) Once
(b) 7 times
(c) 12 times
(d) Does not span the plasma membrane

268. The following is not true for P-glycoprotein:

(a) An ABC protein
(b) An influx pump
(c) Responsible for multidrug resistance
(d) 2 ATP-binding sites per molecule

269. In the bones, receptor activator of nuclear factor-κB (RANK) is expressed on:

(a) Osteoclasts
(b) Osteoblasts
(c) Stromal cells other than osteoblasts
(d) Tumor cells

270. The following is true for osteoprotegerin (OPG):
(a) A decoy receptor for RANK-ligand
(b) Inhibits RANK activation
(c) Intravenous recombinant OPG prevents bone destruction in multiple myeloma
(d) All the above

271. The following is not true for bevacizumab:
(a) A monoclonal antibody
(b) Inhibits vascular endothelial growth factor (VEGF)
(c) Effective in renal cell carcinoma
(d) Hypotension is the major side effect

272. The following is anti-angiogenic:
(a) Thrombospondin-1
(b) Endostatin
(c) Tumstatin
(d) All the above

273. Pemetrexed inhibits:
(a) Thymidylate synthetase
(b) Dihydrofolate reductase
(c) Glycinamide ribonucleotide formyl transferase
(d) All the above

274. The anticancer drug 2-deoxycoformycin acts by:
(a) Inhibiting adenosine deaminase
(b) Inhibiting ribonucleotide reductase
(c) Microtubule depolymerization and disassembly
(d) Microtubule stabilization but with loss of function

275. The anticancer drug hydroxyurea acts by:
(a) Inhibiting adenosine deaminase
(b) Inhibiting ribonucleotide reductase
(c) Microtubule depolymerization and disassembly
(d) Microtubule stabilization but with loss of function

276. The anticancer drug vincristine acts by:
(a) Inhibiting adenosine deaminase
(b) Inhibiting ribonucleotide reductase

(c) Microtubule depolymerization and disassembly
(d) Microtubule stabilization but with loss of function

277. The anticancer drug paclitaxel acts by:
(a) Inhibiting adenosine deaminase
(b) Inhibiting ribonucleotide reductase
(c) Microtubule depolymerization and disassembly
(d) Microtubule stabilization but with loss of function

278. The anticancer drug imatinib inhibits:
(a) Tyrosine kinase (b) Proteasome
(c) Histone deacetylase (d) DNA methyl transferase

279. The anticancer drug bortezomib inhibits:
(a) Tyrosine kinase (b) Proteasome
(c) Histone deacetylase (d) DNA methyl transferase

280. The anticancer drug vorinostat inhibits:
(a) Tyrosine kinase (b) Proteasome
(c) Histone deacetylase (d) DNA methyl transferase

281. The anticancer drug 5-azacytidine inhibits:
(a) Tyrosine kinase (b) Proteasome
(c) Histone deacetylase (d) DNA methyl transferase

282. The following indicates a worse prognosis in breast cancer:
(a) p53 mutation
(b) Lack of estrogen and progesterone receptors
(c) Overexpression of her-2/neu
(d) All the above

283. The positivity rate of estrogen receptors in male breast cancer is:
(a) 0% (b) 10%
(c) 50% (d) 90%

284. The following is not true for serum des-γ-carboxy prothrombin:
(a) Increased in hepatocellular carcinoma
(b) Increased with warfarin therapy

(c) Increased following vitamin K overdose
(d) Indicates immaturation of prothrombin

285. Hepatocellular carcinoma may be inhibited by high dose of vitamin:

(a) A
(b) D
(c) E
(d) K

286. The following seromarker is specific for fibrolamellar hepatocellular carcinoma:

(a) Neurotensin
(b) CEA
(c) AFP
(d) Des-γ-carboxy prothrombin

287. Prostate specific antigen (PSA) is a:

(a) Receptor tyrosine kinase
(b) Kallikrein-like serine protease
(c) Threonine phosphatase
(d) Cyclin

288. The following cytokeratin is a useful tumor marker to establish the site of tumor origin:

(a) CK7
(b) CK20
(c) Both
(d) None of the above

289. The following hormone may be ectopically produced in paraneoplastic syndrome:

(a) PTHrP
(b) Calcitriol
(c) Vasopressin
(d) All the above

290. The following is a phosphatonin:

(a) Fibroblast growth factor (FGF23)
(b) Calcitonin
(c) TNF-α
(d) Calcitriol

291. The following is not true for phosphatonin:

(a) FGF23 is a good example

(b) Increased in oncogenic osteomalacia, a paraneoplastic syndrome
(c) Inhibits renal tubular phosphate reabsorption
(d) Stimulates renal conversion of 25-OH-D_3 to 1,25-$(OH)_2$-D_3

292. The following are not paraneoplastic anti-neuronal antibodies:

(a) Antibodies against acetyl choline receptors
(b) Antibodies against voltage-gated Ca^{2+}-channels
(c) Antibodies against voltage-gated K^+-channels
(d) Anti-neutrophil cytoplasmic antibodies (ANCA)

293. The Jak-Stat signal transduction pathway regulates:

(a) Hemopoiesis (b) Inflammatory response
(c) Immune response (d) All the above

294. Aberrant Jak-Stat pathway is implicated in:

(a) Psoriasis (b) Polycythemia vera
(c) Rheumatoid arthritis (d) All the above

295. The following is true for aristolochic acid:

(a) Implicated in Balkan nephropathy
(b) Forms aristolactam-DNA adducts
(c) May be mutagenic
(d) All the above

296. The following is true for Tamm-Horsfall protein:

(a) Encoded by uromodulin gene (*UMOD*)
(b) Mutation results in familial juvenile hyperuricemic nephropathy
(c) Overlapping features with medullary cystic kidney disease type 2
(d) All of the above

297. The following is true for presenilin:

(a) A chaperone
(b) Regulates apoptosis

(c) Mutated presenilin contributes to aggressive early-onset familial Alzheimer's disease
(d) All the above

298. The following is true for synucleins:
(a) Found in synaptic vesicles
(b) Chaperones
(c) Mutated synucleins contribute to Parkinson's disease
(d) All the above

299. Gene function can be understood by:
(a) Forward genetics (b) Reverse genetics
(c) Both (d) None of the above

300. Most cancer cells have a rapid rate of anaerobic glycolysis, a phenomenon called:
(a) Warburg effect (b) Pasteur effect
(c) Bohr effect (d) CREB-CRE effect

Answers

SECTION-1

1. (c)	2. (a)	3. (a)	4. (c)	5. (d)	6. (d)	7. (c)
8. (a)	9. (a)	10. (c)	11. (b)	12. (c)	13. (d)	14. (a)
15. (a)	16. (b)	17. (d)	18. (a)	19. (b)	20. (d)	21. (a)
22. (c)	23. (a)	24. (b)	25. (b)	26. (a)	27. (c)	28. (d)
29. (d)	30. (a)	31. (c)	32. (c)	33. (b)	34. (d)	35. (d)
36. (a)	37. (d)	38. (a)	39. (c)	40. (a)	41. (c)	42. (b)
43. (b)	44. (a)	45. (c)	46. (a)	47. (d)	48. (d)	49. (d)
50. (b)	51. (d)	52. (b)	53. (a)	54. (d)	55. (c)	56. (a)
57. (d)	58. (b)	59. (d)	60. (a)	61. (a)	62. (d)	63. (c)
64. (d)	65. (d)	66. (d)	67. (c)	68. (a)	69. (d)	70. (b)
71. (d)	72. (a)	73. (c)	74. (d)	75. (a)	76. (a)	77. (a)
78. (b)	79. (c)	80. (d)	81. (a)	82. (b)	83. (c)	84. (d)
85. (a)	86. (c)	87. (b)	88. (c)	89. (a)	90. (b)	91. (a)
92. (d)	93. (a)	94. (b)	95. (d)	96. (d)	97. (c)	98. (d)
99. (a)	100. (b)					

SECTION-2

1. (c)	2. (b)	3. (c)	4. (c)	5. (d)	6. (a)	7. (c)
8. (a)	9. (d)	10. (a)	11. (b)	12. (a)	13. (a)	14. (c)

15. (d) 16. (a) 17. (a) 18. (d) 19. (a) 20. (c) 21. (a)
22. (c) 23. (a) 24. (d) 25. (c) 26. (a) 27. (a) 28. (d)
29. (a) 30. (c) 31. (d) 32. (d) 33. (c) 34. (a) 35. (c)
36. (d) 37. (d) 38. (d) 39. (a) 40. (a) 41. (a) 42. (c)
43. (a) 44. (d) 45. (d) 46. (d) 47. (d) 48. (b) 49. (b)
50. (d) 51. (b) 52. (d) 53. (c) 54. (c) 55. (b) 56. (d)
57. (c) 58. (c) 59. (a) 60. (d) 61. (d) 62. (a) 63. (c)
64. (d) 65. (a) 66. (b) 67. (c) 68. (d) 69. (a) 70. (d)
71. (c) 72. (c) 73. (d) 74. (d) 75. (c) 76. (d) 77. (d)
78. (a) 79. (d) 80. (a) 81. (a) 82. (a) 83. (d) 84. (b)
85. (c) 86. (c) 87. (a) 88. (c) 89. (a) 90. (a) 91. (b)
92. (c) 93. (a) 94. (d) 95. (b) 96. (c) 97. (c) 98. (d)
99. (b) 100. (a) 101. (b) 102. (b) 103. (a) 104. (c) 105. (c)
106. (d) 107. (a) 108. (b) 109. (a) 110. (b) 111. (c) 112. (d)
113. (c) 114. (b) 115. (d) 116. (d) 117. (d) 118. (d) 119. (a)
120. (c) 121. (b) 122. (d) 123. (a) 124. (c) 125. (a) 126. (b)
127. (c) 128. (b) 129. (d) 130. (b) 131. (b) 132. (a) 133. (d)
134. (a) 135. (c) 136. (a) 137. (b) 138. (d) 139. (c) 140. (c)
141. (d) 142. (a) 143. (b) 144. (c) 145. (b) 146. (a) 147. (d)
148. (b) 149. (a) 150. (b) 151. (d) 152. (b) 153. (a) 154. (b)
155. (b) 156. (a) 157. (d) 158. (c) 159. (c) 160. (a) 161. (b)
162. (a) 163. (c) 164. (d) 165. (b) 166. (c) 167. (b) 168. (a)
169. (b) 170. (d) 171. (a) 172. (c) 173. (d) 174. (a) 175. (d)
176. (c) 177. (a) 178. (c) 179. (d) 180. (a) 181. (b) 182. (b)
183. (a) 184. (c) 185. (c) 186. (d) 187. (a) 188. (c) 189. (d)
190. (d) 191. (a) 192. (b) 193. (d) 194. (a) 195. (c) 196. (a)
197. (b) 198. (c) 199. (c) 200. (a) 201. (a) 202. (d) 203. (b)

204. (d)	205. (c)	206. (b)	207. (a)	208. (c)	209. (a)	210. (c)
211. (d)	212. (a)	213. (a)	214. (b)	215. (c)	216. (b)	217. (c)
218. (d)	219. (a)	220. (b)	221. (b)	222. (b)	223. (b)	224. (a)
225. (d)	226. (c)	227. (c)	228. (c)	229. (c)	230. (a)	231. (a)
232. (b)	233. (c)	234. (d)	235. (a)	236. (c)	237. (b)	238. (a)
239. (c)	240. (d)	241. (a)	242. (d)	243. (a)	244. (b)	245. (d)
246. (c)	247. (a)	248. (d)	249. (d)	250. (b)	251. (a)	252. (a)
253. (a)	254. (b)	255. (a)	256. (c)	257. (b)	258. (a)	259. (d)
260. (d)	261. (a)	262. (c)	263. (b)	264. (a)	265. (b)	266. (c)
267. (d)	268. (d)	269. (d)	270. (b)	271. (d)	272. (d)	273. (d)
274. (d)	275. (c)	276. (b)	277. (d)	278. (d)	279. (a)	280. (b)
281. (c)	282. (a)	283. (b)	284. (d)	285. (c)	286. (a)	287. (b)
288. (a)	289. (d)	290. (c)	291. (a)	292. (b)	293. (d)	294. (c)
295. (b)	296. (d)	297. (c)	298. (a)	299. (d)	300. (d)	301. (b)
302. (c)	303. (b)	304. (c)	305. (d)	306. (c)	307. (a)	308. (d)
309. (d)	310. (a)	311. (d)	312. (d)	313. (d)	314. (c)	315. (b)
316. (c)	317. (c)	318. (c)	319. (c)	320. (d)	321. (d)	322. (d)
323. (a)	324. (d)	325. (d)	326. (d)	327. (d)	328. (a)	329. (a)
330. (d)	331. (a)	332. (a)	333. (c)	334. (c)	335. (d)	336. (a)
337. (b)	338. (d)	339. (d)	340. (c)	341. (d)	342. (a)	343. (c)
344. (a)	345. (a)	346. (a)	347. (c)	348. (a)	349. (d)	350. (b)
351. (a)	352. (d)	353. (c)	354. (d)	355. (c)	356. (a)	357. (c)
358. (b)	359. (d)	360. (c)	361. (a)	362. (a)	363. (b)	364. (d)
365. (d)	366. (c)	367. (c)	368. (c)	369. (d)	370. (d)	371. (c)
372. (c)	373. (a)	374. (a)	375. (d)	376. (b)	377. (d)	378. (d)
379. (c)	380. (d)	381. (a)	382. (b)	383. (d)	384. (c)	385. (d)
386. (a)	387. (b)	388. (d)	389. (b)	390. (b)	391. (c)	392. (d)

393. (b) 394. (c) 395. (a) 396. (b) 397. (d) 398. (d) 399. (b)
400. (b) 401. (c) 402. (a) 403. (c) 404. (d) 405. (c) 406. (d)
407. (b) 408. (b) 409. (a) 410. (b) 411. (a) 412. (d) 413. (a)
414. (b) 415. (d) 416. (d) 417. (b) 418. (d) 419. (d) 420. (a)
421. (d) 422. (c) 423. (a) 424. (b) 425. (a) 426. (b) 427. (c)
428. (d) 429. (c) 430. (d) 431. (c) 432. (d) 433. (d) 434. (d)
435. (d) 436. (a) 437. (a) 438. (b) 439. (c) 440. (d) 441. (a)
442. (c) 443. (a) 444. (c) 445. (b) 446. (b) 447. (b) 448. (d)
449. (d) 450. (b) 451. (b) 452. (d) 453. (d) 454. (c) 455. (d)
456. (d) 457. (c) 458. (b) 459. (d) 460. (b) 461. (d) 462. (b)
463. (c) 464. (d) 465. (b) 466. (a) 467. (b) 468. (a) 469. (c)
470. (d) 471. (a) 472. (d) 473. (a) 474. (b) 475. (b) 476. (d)
477. (b) 478. (c) 479. (b) 480. (a) 481. (c) 482. (b) 483. (c)
484. (d) 485. (b) 486. (d) 487. (b) 488. (b) 489. (b) 490. (c)
491. (c) 492. (a) 493. (a) 494. (b) 495. (c) 496. (d) 497. (a)
498. (c) 499. (a) 500. (c) 501. (b) 502. (c) 503. (a) 504. (b)
505. (d) 506. (a) 507. (c) 508. (d) 509. (c) 510. (a) 511. (c)
512. (a) 513. (a) 514. (b) 515. (d) 516. (a) 517. (d) 518. (d)
519. (b) 520. (d) 521. (d) 522. (c) 523. (b) 524. (c) 525. (c)
526. (d) 527. (d) 528. (a) 529. (c) 530. (c) 531. (c) 532. (d)
533. (d) 534. (b) 535. (b) 536. (b) 537. (a) 538. (c) 539. (a)
540. (c) 541. (c) 542. (b) 543. (b) 544. (c) 545. (a) 546. (c)
547. (d) 548. (c) 549. (d) 550. (d) 551. (a) 552. (c) 553. (c)
554. (a) 555. (b) 556. (c) 557. (d) 558. (d) 559. (c) 560. (a)
561. (c) 562. (c) 563. (c) 564. (c) 565. (c) 566. (b) 567. (d)
568. (d) 569. (b) 570. (c) 571. (d) 572. (d) 573. (c) 574. (c)
575. (d) 576. (d) 577. (d) 578. (c) 579. (b) 580. (b) 581. (d)

582. (a)	583. (b)	584. (d)	585. (d)	586. (a)	587. (b)	588. (d)
589. (d)	590. (d)	591. (a)	592. (a)	593. (b)	594. (c)	595. (d)
596. (a)	597. (c)	598. (d)	599. (b)	600. (d)		

SECTION-3

1. (c)	2. (d)	3. (a)	4. (b)	5. (a)	6. (b)	7. (a)
8. (d)	9. (d)	10. (a)	11. (d)	12. (a)	13. (a)	14. (b)
15. (c)	16. (b)	17. (c)	18. (d)	19. (b)	20. (b)	21. (c)
22. (d)	23. (c)	24. (c)	25. (c)	26. (d)	27. (c)	28. (a)
29. (a)	30. (b)	31. (c)	32. (a)	33. (b)	34. (c)	35. (c)
36. (c)	37. (c)	38. (a)	39. (d)	40. (d)	41. (c)	42. (a)
43. (d)	44. (a)	45. (b)	46. (b)	47. (c)	48. (b)	49. (c)
50. (c)	51. (a)	52. (b)	53. (a)	54. (c)	55. (c)	56. (b)
57. (a)	58. (d)	59. (a)	60. (d)	61. (d)	62. (c)	63. (b)
64. (b)	65. (c)	66. (d)	67. (c)	68. (c)	69. (b)	70. (b)
71. (a)	72. (b)	73. (a)	74. (c)	75. (b)	76. (b)	77. (b)
78. (b)	79. (c)	80. (d)	81. (b)	82. (a)	83. (c)	84. (c)
85. (a)	86. (d)	87. (a)	88. (b)	89. (a)	90. (c)	91. (a)
92. (c)	93. (b)	94. (a)	95. (c)	96. (c)	97. (d)	98. (c)
99. (d)	100. (d)	101. (a)	102. (c)	103. (d)	104. (d)	105. (b)
106. (b)	107. (d)	108. (c)	109. (a)	110. (d)	111. (d)	112. (b)
113. (b)	114. (c)	115. (d)	116. (a)	117. (d)	118. (c)	119. (c)
120. (c)	121. (c)	122. (c)	123. (b)	124. (c)	125. (c)	126. (a)
127. (b)	128. (d)	129. (d)	130. (d)	131. (d)	132. (c)	133. (c)
134. (d)	135. (d)	136. (c)	137. (d)	138. (d)	139. (c)	140. (a)
141. (c)	142. (b)	143. (b)	144. (a)	145. (a)	146. (d)	147. (b)
148. (a)	149. (d)	150. (b)	151. (a)	152. (c)	153. (d)	154. (c)

155. (d) 156. (a) 157. (b) 158. (c) 159. (c) 160. (c) 161. (b)
162. (d) 163. (d) 164. (d) 165. (a) 166. (a) 167. (a) 168. (b)
169. (c) 170. (a) 171. (a) 172. (c) 173. (d) 174. (b) 175. (a)
176. (b) 177. (d) 178. (b) 179. (a) 180. (d) 181. (d) 182. (c)
183. (b) 184. (b) 185. (c) 186. (b) 187. (d) 188. (c) 189. (b)
190. (d) 191. (b) 192. (a) 193. (b) 194. (c) 195. (c) 196. (a)
197. (a) 198. (b) 199. (a) 200. (d)

SECTION-4

1. (a) 2. (b) 3. (b) 4. (b) 5. (c) 6. (b) 7. (c)
8. (c) 9. (a) 10. (b) 11. (c) 12. (d) 13. (d) 14. (a)
15. (b) 16. (c) 17. (b) 18. (c) 19. (d) 20. (d) 21. (b)
22. (a) 23. (d) 24. (c) 25. (d) 26. (a) 27. (d) 28. (a)
29. (c) 30. (b) 31. (d) 32. (b) 33. (d) 34. (d) 35. (a)
36. (b) 37. (d) 38. (c) 39. (a) 40. (b) 41. (d) 42. (b)
43. (d) 44. (a) 45. (c) 46. (a) 47. (d) 48. (a) 49. (b)
50. (a) 51. (c) 52. (d) 53. (a) 54. (d) 55. (d) 56. (d)
57. (b) 58. (a) 59. (d) 60. (a) 61. (d) 62. (c) 63. (d)
64. (d) 65. (a) 66. (b) 67. (d) 68. (a) 69. (d) 70. (a)
71. (a) 72. (c) 73. (b) 74. (a) 75. (a) 76. (d) 77. (d)
78. (d) 79. (c) 80. (b) 81. (d) 82. (c) 83. (a) 84. (b)
85. (d) 86. (a) 87. (d) 88. (a) 89. (b) 90. (c) 91. (d)
92. (b) 93. (c) 94. (a) 95. (d) 96. (d) 97. (a) 98. (c)
99. (a) 100. (b) 101. (a) 102. (d) 103. (b) 104. (d) 105. (c)
106. (d) 107. (c) 108. (b) 109. (b) 110. (c) 111. (a) 112. (a)
113. (a) 114. (d) 115. (b) 116. (d) 117. (c) 118. (a) 119. (d)
120. (d) 121. (c) 122. (a) 123. (b) 124. (b) 125. (a) 126. (c)

127. (b) 128. (d) 129. (c) 130. (a) 131. (c) 132. (d) 133. (c)
134. (b) 135. (b) 136. (d) 137. (a) 138. (d) 139. (d) 140. (a)
141. (c) 142. (a) 143. (b) 144. (c) 145. (d) 146. (b) 147. (a)
148. (c) 149. (d) 150. (d) 151. (c) 152. (a) 153. (a) 154. (d)
155. (b) 156. (c) 157. (b) 158. (d) 159. (b) 160. (d) 161. (b)
162. (a) 163. (c) 164. (d) 165. (a) 166. (a) 167. (a) 168. (b)
169. (a) 170. (b) 171. (a) 172. (d) 173. (c) 174. (a) 175. (d)
176. (c) 177. (c) 178. (a) 179. (d) 180. (c) 181. (d) 182. (a)
183. (c) 184. (c) 185. (a) 186. (d) 187. (b) 188. (d) 189. (d)
190. (a) 191. (a) 192. (d) 193. (d) 194. (c) 195. (d) 196. (a)
197. (b) 198. (b) 199. (d) 200. (a) 201. (a) 202. (a) 203. (b)
204. (a) 205. (a) 206. (a) 207. (c) 208. (d) 209. (d) 210. (d)
211. (d) 212. (d) 213. (a) 214. (a) 215. (d) 216. (a) 217. (b)
218. (d) 219. (c) 220. (a) 221. (d) 222. (c) 223. (b) 224. (a)
225. (a) 226. (d) 227. (d) 228. (a) 229. (b) 230. (c) 231. (c)
232. (a) 233. (d) 234. (d) 235. (b) 236. (a) 237. (d) 238. (a)
239. (a) 240. (b) 241. (c) 242. (a) 243. (a) 244. (b) 245. (d)
246. (d) 247. (a) 248. (d) 249. (a) 250. (b) 251. (b) 252. (c)
253. (b) 254. (b) 255. (b) 256. (d) 257. (d) 258. (c) 259. (d)
260. (d) 261. (b) 262. (d) 263. (d) 264. (c) 265. (a) 266. (c)
267. (a) 268. (b) 269. (d) 270. (c) 271. (a) 272. (b) 273. (b)
274. (c) 275. (d) 276. (a) 277. (c) 278. (b) 279. (a) 280. (d)
281. (c) 282. (d) 283. (d) 284. (d) 285. (c) 286. (d) 287. (a)
288. (b) 289. (d) 290. (a) 291. (d) 292. (d) 293. (b) 294. (a)
295. (d) 296. (c) 297. (d) 298. (c) 299. (a) 300. (a)

SECTION-5

1. (a) 2. (b) 3. (d) 4. (c) 5. (b) 6. (c) 7. (c)
8. (d) 9. (b) 10. (c) 11. (c) 12. (d) 13. (c) 14. (a)

15. (b)	16. (d)	17. (d)	18. (d)	19. (a)	20. (d)	21. (c)
22. (b)	23. (c)	24. (d)	25. (b)	26. (c)	27. (a)	28. (a)
29. (b)	30. (a)	31. (d)	32. (c)	33. (c)	34. (a)	35. (a)
36. (d)	37. (c)	38. (a)	39. (d)	40. (c)	41. (c)	42. (d)
43. (a)	44. (b)	45. (c)	46. (d)	47. (a)	48. (b)	49. (b)
50. (d)	51. (a)	52. (b)	53. (b)	54. (d)	55. (d)	56. (d)
57. (d)	58. (a)	59. (c)	60. (c)	61. (d)	62. (c)	63. (a)
64. (b)	65. (a)	66. (a)	67. (d)	68. (a)	69. (d)	70. (c)
71. (b)	72. (c)	73. (c)	74. (d)	75. (d)	76. (c)	77. (c)
78. (d)	79. (d)	80. (a)	81. (a)	82. (d)	83. (d)	84. (d)
85. (d)	86. (d)	87. (c)	88. (d)	89. (d)	90. (d)	91. (d)
92. (d)	93. (a)	94. (c)	95. (a)	96. (a)	97. (b)	98. (b)
99. (a)	100. (d)	101. (d)	102. (c)	103. (a)	104. (b)	105. (a)
106. (b)	107. (c)	108. (d)	109. (b)	110. (a)	111. (c)	112. (c)
113. (d)	114. (d)	115. (c)	116. (d)	117. (a)	118. (c)	119. (a)
120. (c)	121. (d)	122. (d)	123. (d)	124. (a)	125. (d)	126. (b)
127. (a)	128. (c)	129. (a)	130. (b)	131. (d)	132. (a)	133. (d)
134. (d)	135. (c)	136. (d)	137. (d)	138. (b)	139. (b)	140. (b)
141. (a)	142. (c)	143. (d)	144. (a)	145. (c)	146. (a)	147. (a)
148. (a)	149. (d)	150. (a)	151. (d)	152. (a)	153. (c)	154. (b)
155. (b)	156. (a)	157. (d)	158. (c)	159. (a)	160. (c)	161. (a)
162. (a)	163. (a)	164. (b)	165. (d)	166. (c)	167. (b)	168. (c)
169. (a)	170. (b)	171. (d)	172. (d)	173. (a)	174. (d)	175. (d)
176. (a)	177. (b)	178. (d)	179. (b)	180. (a)	181. (a)	182. (a)
183. (a)	184. (a)	185. (c)	186. (d)	187. (c)	188. (d)	189. (a)
190. (b)	191. (c)	192. (b)	193. (a)	194. (c)	195. (d)	196. (d)
197. (d)	198. (b)	199. (b)	200. (b)			

SECTION-6

1. (c)	2. (d)	3. (a)	4. (c)	5. (c)	6. (c)	7. (b)
8. (d)	9. (a)	10. (c)	11. (a)	12. (a)	13. (b)	14. (a)
15. (b)	16. (c)	17. (d)	18. (b)	19. (c)	20. (d)	21. (b)
22. (d)	23. (c)	24. (d)	25. (d)	26. (a)	27. (a)	28. (c)
29. (c)	30. (d)	31. (b)	32. (b)	33. (c)	34. (a)	35. (c)
36. (b)	37. (c)	38. (b)	39. (b)	40. (d)	41. (d)	42. (b)
43. (d)	44. (d)	45. (a)	46. (d)	47. (c)	48. (c)	49. (b)
50. (c)	51. (c)	52. (a)	53. (c)	54. (c)	55. (c)	56. (c)
57. (d)	58. (c)	59. (a)	60. (b)	61. (d)	62. (d)	63. (d)
64. (d)	65. (c)	66. (d)	67. (a)	68. (d)	69. (a)	70. (b)
71. (b)	72. (a)	73. (c)	74. (d)	75. (d)	76. (c)	77. (d)
78. (d)	79. (d)	80. (c)	81. (c)	82. (d)	83. (d)	84. (a)
85. (d)	86. (d)	87. (d)	88. (c)	89. (d)	90. (a)	91. (c)
92. (d)	93. (c)	94. (d)	95. (a)	96. (d)	97. (d)	98. (a)
99. (c)	100. (b)	101. (a)	102. (b)	103. (d)	104. (b)	105. (a)
106. (b)	107. (d)	108. (a)	109. (c)	110. (c)	111. (c)	112. (b)
113. (a)	114. (b)	115. (c)	116. (b)	117. (c)	118. (a)	119. (b)
120. (a)	121. (d)	122. (a)	123. (c)	124. (a)	125. (a)	126. (d)
127. (a)	128. (a)	129. (b)	130. (c)	131. (b)	132. (d)	133. (b)
134. (c)	135. (a)	136. (b)	137. (c)	138. (a)	139. (d)	140. (d)
141. (d)	142. (b)	143. (c)	144. (a)	145. (c)	146. (c)	147. (d)
148. (a)	149. (a)	150. (a)	151. (d)	152. (d)	153. (d)	154. (d)
155. (d)	156. (c)	157. (b)	158. (b)	159. (d)	160. (c)	161. (c)
162. (c)	163. (b)	164. (a)	165. (d)	166. (b)	167. (d)	168. (a)
169. (c)	170. (d)	171. (d)	172. (c)	173. (a)	174. (c)	175. (d)
176. (c)	177. (c)	178. (b)	179. (c)	180. (b)	181. (a)	182. (b)
183. (c)	184. (d)	185. (c)	186. (c)	187. (d)	188. (d)	189. (d)
190. (a)	191. (d)	192. (b)	193. (a)	194. (d)	195. (c)	196. (d)

197. (d)	198. (c)	199. (d)	200. (d)	201. (c)	202. (d)	203. (c)
204. (a)	205. (b)	206. (c)	207. (d)	208. (b)	209. (d)	210. (a)
211. (d)	212. (d)	213. (c)	214. (c)	215. (b)	216. (d)	217. (d)
218. (b)	219. (a)	220. (c)	221. (d)	222. (b)	223. (d)	224. (d)
225. (a)	226. (d)	227. (c)	228. (a)	229. (d)	230. (c)	231. (b)
232. (a)	233. (d)	234. (a)	235. (c)	236. (d)	237. (d)	238. (b)
239. (c)	240. (d)	241. (a)	242. (a)	243. (b)	244. (b)	245. (a)
246. (d)	247. (b)	248. (a)	249. (c)	250. (c)	251. (d)	252. (a)
253. (a)	254. (c)	255. (a)	256. (c)	257. (d)	258. (c)	259. (a)
260. (a)	261. (d)	262. (d)	263. (d)	264. (c)	265. (c)	266. (c)
267. (c)	268. (b)	269. (a)	270. (d)	271. (d)	272. (d)	273. (d)
274. (a)	275. (b)	276. (c)	277. (d)	278. (a)	279. (b)	280. (c)
281. (d)	282. (d)	283. (d)	284. (c)	285. (d)	286. (a)	287. (b)
288. (c)	289. (d)	290. (a)	291. (d)	292. (d)	293. (d)	294. (d)
295. (d)	296. (d)	297. (d)	298. (d)	299. (d)	300. (a)	

References

Michael M. Cox, David L. Nelson, "Lehninger Principles of Biochemistry", 5th edition, 2010, W.H. Freeman and Company, New York.

Jeremy M. Berg, John L. Tymoczko, Lubert Stryer, "Biochemistry", 6th edition, 2007, W.H. Freeman and Company, New York.

Bruce Alberts, Alexander Johnson, Julian Lewis, Martin Raff, Peter Walter, "Molecular Biology of the Cell", 4th edition, 2002, Garland Science (Taylor and Francis), New York.

Thomas M. Devlin, "Textbook of Biochemistry with Clinical Correlations", 6th edition, 2006, Wiley-Liss, New Jersey.

Henry M. Kronenberg, Shlomo Melmed, Kenneth S. Polonsky, P. Reed Larsen, "Williams Textbook of Endocrinology, 11th edition, 2008, Saunders (Elsevier), Philadelphia.

Jocelyn E. Krebs, Elliott S. Goldstein, Stephen T. Kilpatrick, "Lewin's Genes X", 10th edition, 2011, Jones and Bartlett Publishers, Sudbury.

Lynn B. Jorde, John, C. Carey, Michael J. Bamshad, "Medical Genetics", 4th edition, 2010, Mosby (Elsevier), Philadelphia.

Dan L. Longo, Anthony S. Fauci, Dennis L. Kasper, Stephen L. Hauser, J. Larry Jameson, Joseph Loscalzo, "Harrison's Principles of Internal Medicine", 18th edition, 2012, McGraw Hill, New York.

Robert K. Murray, Daryl K. Granner, Peter A. Mayes, Victor W. Rodwell, "Harper's Biochemistry, 24th edition, 1998, Prentice-Hall International, Inc., Singapore.

Frances Fischbach, Marshall B. Dunning III, "A Manual of Laboratory and Diagnostic Tests", 8th edition, 2009, Wolters Kluwer/Lippincott Williams and Wilkins, New Delhi.

E.N. Whitney, S.R. Rolfes, "Understanding Nutrition", 8th edition, 1999, An International Thomson Publishing Company, Belmont.

Carl A. Burtis, Edward R. Ashwood, "Tietz Textbook of Clinical Chemistry", 3rd edition, 1999, W.B. Saunders Company (Indian Edition), New Delhi.

D.M. Vasudevan, SreeKumari S, Kannan Vaidyanathan, "Textbook of Biochemistry for Medical Students, 6th edition, 2011, Jaypee Brothers Medical Publishers (P) Ltd., Kochi.

Harbans Lal, Rajesh Pandey, "Textbook of Biochemistry", 2nd edition, 2011, CBS Publishers & Distributors Pvt. Ltd., New Delhi.